Constructivity and Computability in Historical and Philosophical Perspective

LOGIC, EPISTEMOLOGY, AND THE UNITY OF SCIENCE

VOLUME 34

Logic, Epistemology, and the Unity of Science aims to reconsider the question of the unity of science in light of recent developments in logic. At present, no single logical, semantical or methodological framework dominates the philosophy of science. However, the editors of this series believe that formal techniques like, for example, independence friendly logic, dialogical logics, multimodal logics, game theoretic semantics and linear logics, have the potential to cast new light on basic issues in the discussion of the unity of science.

This series provides a venue where philosophers and logicians can apply specific technical insights to fundamental philosophical problems. While the series is open to a wide variety of perspectives, including the study and analysis of argumentation and the critical discussion of the relationship between logic and the philosophy of science, the aim is to provide an integrated picture of the scientific enterprise in all its diversity.

More information about this series at http://www.springer.com/series/6936

Jacques Dubucs • Michel Bourdeau
Editors

Constructivity and Computability in Historical and Philosophical Perspective

Springer

Editors
Jacques Dubucs
Michel Bourdeau
Institut d'histoire et de philosophie
des sciences et des techniques
(IHPST; CNRS-Paris1-ENS)
Paris, France

ISSN 2214-9775 ISSN 2214-9783 (electronic)
ISBN 978-94-017-9216-5 ISBN 978-94-017-9217-2 (eBook)
DOI 10.1007/978-94-017-9217-2
Springer Dordrecht Heidelberg New York London

Library of Congress Control Number: 2014947463

Printed on acid-free paper

Springer is part of Springer Science+Business Media (www.springer.com)

Preface

This book covers the almost 80 years from Turing's seminal paper of 1936 to the present and focuses on two developments during that period.[1]

On the one hand, the development of computability theory and complexity theory called for generalizations, restrictions, and other modifications of Turing's original machine. On the other hand, it became clear that recursion theory cannot serve as a foundational analysis of the notion of computable function(al), which has to be accepted as primitive.

In different ways, both of these developments contributed to a unified view of logic. The first development helped bringing various phenomena together into a

[1]In her recent book *Logic and Philosophy of Mathematics in the Early Husserl* (Dordrecht: Springer 2010), Stefania Centrone has drawn attention to the fact that the first to call for a theoretical study of computability probably was Husserl. In his *Philosophy of Arithmetic* of 1891 (*Collected Works* vol. X, trl. D. Willard, Dordrecht: Springer 2003), he formulated the following 'general postulate of arithmetic': 'the symbolic formations that are different from the systematic numbers must, wherever they turn up, be reduced to the systematic numbers equivalent to them, as their normative forms. Accordingly there arises, as the first task of Arithmetic, to separate all conceivable symbolic modes of formation of numbers into their distinct types, and to discover for each type the methods that are reliable and as simple as possible for carrying out that reduction' (p. 277). He considered the arithmetical operations to be but the methods to carry out that reduction (p. 277), and understood computation, in arithmetic but also more generally, as 'any rule-governed mode of derivation of signs from signs within any algorithmic sign-system according to the "laws"—or better: the conventions—for combination, separation, and transformation peculiar to that system' (p. 273). He also raised the question of the computability of numbers that are defined by a system of equations (pp. 296–298). Husserl did not, however, attempt to develop such a theory of computability, and his suggestion seems not to have been picked up on by his contemporaries either. Moreover, it seems that none of those who came to play a role in the development of computability theory was aware of Husserl's suggestion. They had, of course, ample independent motivation. But it would be interesting to know, for example, whether Skolem knew this chapter by Husserl. Skolem visited Göttingen during the winter of 1915–1916; that was Husserl's last term as a professor there, before moving to Freiburg.

framework in which the original Turing machine turned out to be a special, and in some contexts ideal(ized), case. The second development dispelled a serious misunderstanding according to which recursion theory should ideally replace informal reflection on the notion of computable function. Instead, there is room and indeed need for both, depending on one's purpose.[2]

The two developments are discussed in seven chapters, as follows.

Göran Sundholm's opening chapter, *Constructive Recursive Functions, Church's Thesis, and Brouwer's Theory of the Creating Subject*, first discusses recursive versus constructive functions and, following Heyting, stresses that from a constructive point the former cannot replace the latter. The second half of the paper treats of the Kreisel-Myhill theory CS for Brouwer's Creating Subject, and its relation to BHK meaning-explanations and Kripke's Schema. Kripke's Schema is reformulated as a principle and shown to be classically valid. Assuming existence of a verification-object for this principle, a modification of a proof of conservativeness of Van Dalen's, is shown to give a relative BHK meaning explanation for the Kreisel-Myhill connective. The result offers an explanation of why Kripke's Schema can be used as a replacement of the Theory of Creating Subject when formulating Brouwerian counter-examples. It also shows that the Theory of Creating Subject is classically valid.

The relation between computation and machine is as old as the abacus, but only with Turing's pioneering work to this relation became central to computability theory. Chapter 2, Jean Mosconi's *The Developments of the Concept of Machine Computability from 1936 to the 1960s*, deals with this aspect, which is crucial from a technological point of view. Mosconi explains how Turing's ideas were gradually adopted, developed and modified, leading to something much closer to the actual computer. The development can be divided into three stages. First, there is a strong contrast between a quick acceptance of the conceptual analysis of computation and a scarce use of the technical potentialities of Turing's contribution: While Gödel quickly saw the philosophical relevance of Turing's work (it made him overcome his objections to Church's Thesis), Post was perhaps the first to see its mathematical fruitfulness. The Turing Machine enters in a crucial way in Post's 1947 proof of the algorithmic unsolvability of Thue's problem. Then follows the rise of the theory of Turing Machines in the 1950s. Here the Turing Machine was not used to solve problems or gain philosophical insights, but was taken as a proper object of mathematical investigation. The main idea was to include the Turing Machine in the 'general and logical theory of automata' that Von Neumann suggested to develop in the late 1940s. From this point of view, it became possible to study a larger class of automata: not only finite ones, as it was initially the case with

[2]That leaves open the question whether the notions or recursive function and computable function nevertheless have the same extension. Church's Thesis asserts that they do; but within the intuitionistic theory of the creating subject, Kripke has constructed a computable function that is not recursive. This counterexample is discussed in the papers by Sundholm and Van Atten in the present volume.

Shannon or McCulloch's automata, but also infinite ones. Starting from the general structure of the original Turing Machine, various restrictions and generalizations were defined, leading to a hierarchy of automata, and to a better understanding of what Turing Machines can achieve and how. Finally, the progress of technology made it necessary to rework Turing's model so as to establish the link between computability theory and computer practice. Hao Wang's machine B (1957), which uses the notion of instruction instead of the notion of state, was the first attempt to bridge the gap, but a completely explicit model of a program and register machine was arrived at only 1963, by Stephenson and Sturgis.

With the development of computer science and the increasing demand for feasibility, computability theory gave birth to complexity theory. Chapters 3 and 4, by Serge Grigorieff and Marie Ferbus-Zanda, respectively, are dedicated to Kolmogorov complexity, which may be seen as an offspring of computability theory. Chapter 3, Grigorieff's *Information and randomness*, is the more technical one of the two. Kolmogorov introduced a radically new approach to the measurement of information. In the combinatorial approach, the information content of an object x was defined as the length of the shortest binary word which 'encodes' x. Kolmogorov's idea was to bring in the resources of computability by defining the information content of x as the length of the shortest program which computes x. The chapter gives the main concepts and results, together with their proofs; some related notions of complexity, like Levin monotone complexity or Schnorr process complexity are also presented. A new step was taken when Kolmogorov, and independently Chaitin, noticed that the algorithmic theory of information could also be used to give a definition of randomness, which was still lacking after the axiomatization of probability theory. The basic idea is that a word is random if it is incompressible, that is, if there is no shorter way to describe it. Martin-Löf has shown that the necessary condition was also a sufficient one.

Chapter 4, *Application to Classification Theory*, takes a more philosophical stance and shows the relevance of Kolmogorov complexity to computer science, where it has already found a number of useful applications. With the World Wide Web and its huge network of machines, the analysis of information processing has become even more challenging and the need for a classification even more urgent. Up to now the two main approaches have been classification by compression and the so-called Google classification. But they still lack a good formalisation. As they both use complexity, Ferbus-Zanda proposes to take Kolmogorov algorithmic information theory as the mathematical foundation of information classification we are looking for. The two main approaches can now be formulated in terms of two types of definition of mathematical objects, namely iterative definitions, based on set theoretical union, and inductive or recursive definitions, based on set theoretical intersection; they can also be seen as bottom-up and top-down versions of the same underlying theory, of which Ferbus-Zanda gives a short presentation. Furthermore, she shows how these two dual modes are also found in information systems, particularly the relational database model introduced by Codd in the 1970s.

Chapter 5, *Proof Theoretical Semantics and Feasibility*, by Jean Fichot, returns to constructivity. According to one of the best known justifications of constructive reasoning, the meaning of logical constants is given by their introduction rules or, what amounts to the same, by what counts as a canonical proof of it. But canonical proofs are idealisations of the ones we normally use; depending on what types of knowledge one admits, proof theoretical semantics may therefore be liable to be invalidated by the non-feasibility of the canonical proofs. Fichot's contribution reviews two attempts to overcome this difficulty and to go one step further towards feasible canonical proofs. The first is the new recursion-theoretic characterisation of polytime functions given in 1997 by Bellantoni and Cook. Besides natural numbers, they use feasible numbers and succeed in giving a meaningful explanation in the style of Dummett for those numbers. Unfortunately, there remains a gap between feasible arithmetic and the feasible theory of proofs we are looking for. The second attempt is light affine logic, a system introduced by Asperti in 1998 and further studied by Baillot and Terrui. The main idea comes from light linear logic, where Girard added a new modal operator, '§', to the *of course* operator, '!', of linear logic. Just as '!' allows for contraction, '§' allows for weakening. Besides the formulas that express perennial propositions and can be contracted and reused as often as wanted, and the formulas that are simply true, we must now admit a third kind of formulas. But this splitting of '!' into two modal operators gives very useful tools for controlling the computational complexity of the cut elimination procedure. Jean Fichot explains how the justification for the logical rules of such a system can be given.

The sixth chapter, *Recursive Functions and Constructive Mathematics*, by Thierry Coquand, addresses one of the most fundamental questions concerning the relations between constructivity and computability: is the theory of recursive functions needed for a rigorous development of constructive mathematics? The answer is negative in both the theoretical and the practical sense. The argument proceeds in two steps. The first one shows how the success of recursion theory fostered a lack of sense for constructivity. As was noted early on by Heyting and Skolem, from a constructive point of view, the theory of recursive functions cannot give us a formal definition of the intuitive notion of computable function. Kleene's definition of μ, for instance, uses existential quantification: if $R(x, y)$ is a recursive relation and if $(x)EyR(x, y)$ holds, then $\mu yR(x, y)$ is a recursive function of x. But we are left with a dilemma: if the quantifier is interpreted non-constructively, the relation between computability and constructivity is lost; if it is interpreted constructively, then the definition in fact presupposes some notion of a computable function. The second step in Coquand's argument begins with Bishop's *Foundations of Constructive Analysis* (1967). The book was a breakthrough: It taught us that it was not only theoretically possible but also practically more satisfactory to introduce functions in constructive mathematics without mentioning recursivity. Much current work in constructive mathematics strongly relies on Bishop's ideas. It is also noteworthy that one popular definition of constructive mathematics, according to which it is *mathematics developed using intuitionistic logic*, is independent of any notion of algorithm.

The final chapter, *Gödel and Intuitionism*, by Mark van Atten, starts with a brief survey of Gödel's personal contacts with Brouwer and Heyting. Some examples are discussed where intuitionistic ideas had a direct influence on Gödel's technical work. Then it is argued that the closest rapprochement of Gödel to intuitionism is seen in the development of the Dialectica Interpretation, during which he came to accept the notion of computable functional of finite type as primitive. It is shown that Gödel already thought of that possibility in the Princeton lectures on intuitionism of Spring 1941, and evidence is presented that he adopted it in the same year or the next, long before the publication of 1958. Draft material for the revision of the Dialectica paper is discussed in which Gödel describes the Dialectica Interpretation as being based on a new intuitionistic insight obtained by applying phenomenology, and also notes that relate the new notion of reductive proof to phenomenology. In an appendix, attention is drawn to notes from the archive according to which Gödel anticipated autonomous transfinite progressions when writing his incompleteness paper.

This book has grown out of the meeting 'Computability and Constructivity in Historical and Philosophical Perspective', which took place at the École normale supérieure in Paris, December 17–18, 2006. The organisers were Jacques Dubucs (IHPST), Michel Bourdeau (IHPST), Jean-Paul Delahaye (Université des Sciences et Technologies de Lille), and Gerhard Heinzmann (Université de Nancy II). The meeting was a Joint Session of the two divisions of the International Union of History and Philosophy of Science (IUHPS): the Division of Logic, Methodology and Philosophy of Science (DLMPS) and the Division of History of Science and Technology (DHST).

Paris, France
June 2013

Jacques Dubucs
Michel Bourdeau

Contents

Chapter 1
Constructive Recursive Functions, Church's Thesis, and Brouwer's Theory of the Creating Subject: Afterthoughts on a Parisian Joint Session

Göran Sundholm

Dedicated to Dirk van Dalen on the occasion of his 80th birthday.

'inexorable logic and resolute constructiveness'[1]

Abstract The first half of the paper discusses recursive versus constructive functions and, following Heyting, stresses that from a constructive point the former cannot replace the latter. The second half of the paper treats of the Kreisel-Myhill theory CS for Brouwer's Creating Subject, and its relation to BHK meaning-explanations and Kripke's Schema. Kripke's Schema is reformulated as a principle and shown to be classically valid. Assuming existence of a verification-object for this principle, a modification of a proof of conservativeness of Van Dalen's, is shown to give a relative BHK meaning explanation for the Kreisel-Myhill connective. The result offers an explanation of why Kripke's Schema can be used as a replacement of the Theory of Creating Subject when formulating Brouwerian counter-examples. It also shows that the Theory of Creating Subject is classically valid.

'*Calculemus!*'—the venerable Leibnizian exhortation wants to set us to work at resolving various bones of human contention and strife. In a strict sense, however, only that can be calculated which can be calculated *upon*. *Calculate* (1570),

[1] Stefan Zweig (1979, p. 191), apropos Calvin's *Institutio*.

Originally published at https://www.academia.edu/3528209/Constructive_recursive_functions_Church_thesis_and_Kripkes_schema.

G. Sundholm (✉)
Department of Philosophy, Leiden University, Witte Singel 25, 2311 BZ Leiden, The Netherlands
e-mail: goran.sundholm@gmail.com

J. Dubucs, M. Bourdeau (eds.), *Constructivity and Computability in Historical and Philosophical Perspective*, Logic, Epistemology, and the Unity of Science 34,
DOI 10.1007/978-94-017-9217-2_1

compute (1579), *reckon* (1225) ... the *OED* treats them virtually synonymously, but the Germanic word *reckon* is the oldest. When we reckon we count *on* something: fingers, 'calculi', that is, abacus stones, numerals, signs, ...

Today one often speaks of a calculable (or *computable*) function, without stopping to reflect how strange such language really is, especially from a Platonist ('classical') point of view. There a function $\mathbf{N} \to \mathbf{N}$ is a certain set F, at a fairly low level in the cumulative hierarchy, namely a set of ordered pairs that is unique in the second component, and where every (and only) finite Von Neumann ordinal occurs among the first components. What does it mean for such a thing, object, entity, ... to be *calculable*? It is hard to say. The set F—construed as an object in the cumulative hierarchy—is of course wholly notation-free. There is nothing left to compute on, so to say. It certainly does not mean

$$(**) \quad (\exists e : \mathbf{N})(\forall k : \mathbf{N})(\exists m : \mathbf{N})(T(e, k, n) \& \langle k, U(m) \rangle \in F)$$

where we use Kleene's computation-predicate T and his result-extracting function U.

The claim (**) is true when the function F is (general) recursive, but does not define what it *means* for the set F to be a recursive function, nor was recursiveness the issue here, but *calculability*. The (classical) *truth* of (**) in the cumulative hierarchy is not enough to ensure that an agent is able to perform a calculation; this just ensures that the ontology contains a natural number e with certain properties. For calculability to be guaranteed the calculating agent has to *know* that (**) is true, and also understand the codings involved in order to reconstruct the algorithm from the Kleene-code e. Clearly, not just the function graph (extensionally conceived) is relevant for computability: the way in which the function is actually presented to us plays a decisive role here.

A numerical *expression* might be evaluable by means of a series of steps, might be *calculable*, to a value, but a function, in today's set-theoretic sense, never. An *expression* standing for a certain value of a function might be calculable, might be reckoned out, (might be evaluable) to a numerical value, for instance '$2 + 2$' can be reckoned out to 4. The Riemann hypothesis function, if indeed a function it is, that is presented at (#) below, cannot be evaluated today: we have no way of reckoning out a value for, say, $f(14)$, or indeed for any other argument. The values of a function, when calculable, can perhaps be calculated by means of a certain uniform program or procedure. 'Calculable' is commonly encountered in the combination 'effectively calculable'. Is this use of 'effective' not pleonastic? What use would one have for 'non-effectively calculable'? Would one ever want to say of a function that it is *non-effectively* calculable, and, if so, what would one possible mean by it? A *calculation* could perhaps be called non-effective in a relative way, when the process or method of calculation was very slow. ('It is not very effective, you know ...'.) But this still would not be a non-effective calculation.[2]

[2] I am happy to have Alonzo Church on my side here; see his letter to Yannis Moschovakis that is printed as an appendix. NB. One must not confuse *relative* computability (from an *oracle*) with computability that is not effective. Moschovakis has treated beautifully of relative computability and Church's Thesis in his Heyting lecture at Amsterdam, September 2012.

A calculable function, then, is one whose values can be calculated. Calculability is a matter of agency, of ability to carry out calculations. Since we discuss issues in constructivism our functions are *total*. That f is a function from α to β *means* that one may apply f to any argument in α and get a value. Here the notion of function application is primitive, and, by the meaning explanation for the judgement $f : \alpha \to \beta$, inferences according to the rule

$$\frac{a \in \alpha}{f(a) \in \beta}$$

are valid by stipulation. Accordingly, when I know the judgement $f : \alpha \to \beta$, getting to know $a \in \alpha$ allows me, by stipulation, to proceed to knowledge also of $f(a) \in \beta$.

For simplicity, I confine myself largely to functions from **N** to **N**. Hence $f : \mathbf{N} \to \mathbf{N}$ means that f has a value $f(k) \in \mathbf{N}$, for *every* number $k \in \mathbf{N}$. Kleene (1938) introduced also the notion of a 'partial recursive function' in computability theory. On one reading, this will be a 'partial function that is recursive'. However, strictly speaking, *partial function* is an oxymoron. The adjective 'partial' acts as a *modification* that takes us out of functions, rather than as a qualifying property among functions. A 'partial function' is no function, since it is not defined for every element in the domain. The syntactic form 'partial function' is misleading. Instead, in recursion theory, one could better speak about recursively enumerable functional relations, whether total or not. The reading 'function that is partial recursive', on the other hand, would appear to indicate a (total) function that for some reason is not fully recursive, but only partially so.

So a calculable function is a function $f : \mathbf{N} \to \mathbf{N}$ whose values *can be calculated*. Should the explicit presupposition of agency be taken at face value here? That is, for a function to be calculable, should one be able to *perform* the calculations in question? The agency involved has to be possibility in principle, not practical feasibility. The recursion equations that serve to define Ackermann's function ϕ, which outgrows all primitive recursive functions, clearly offer a *method* of calculation. However, an explicit calculation of the numerical value of, say, $\phi(10, 10, 10)$ will, under our present physical limitations, be beyond human agency.[3]

[3]See Ackermann (1928). My Oxford supervisor Robin Gandy told me that during WWII, at Bletchley Park, Alan Turing proposed a challenging competition:

> What is the largest natural number that can be written on a postcard?

Clearly transfinite hierarchies of fast-growing functions offer a promising way to attack the problem, and hence constructive notation-systems for large ordinals are called for, so as to be able to write terms for very large numbers, though I do not know the answer to Turing's challenge. In correspondence Andrew Hodges pointed out that John von Neumann's proposed something very similar to Stanislaw Ulam already in 1938, in connection with some 'schemata of Turing's'. The matter is dealt with by Hodges (1983, p. 145), and the pertinent letter from Ulam to Hodges can be found at http://www.turing.org.uk/sources/vonneumann.html.

Classical and constructivist mathematicians have different priorities here. Both agree that calculability of a function requires a method of calculation. This method of calculation is given via an algorithm or programme (or, at a further remove, a *code* of such, as in the example $(**)$ above), that is, a collection of procedural instructions according to which calculations are performed. However, a classical—'Platonist'—mathematician does not need to have the algorithm at hand, but will be satisfied also with a classical indirect proof that such an algorithm *exists* in his Platonist ontology. In particular, as noted above, the algorithm in question can be coded as a number, but need not be known as an algorithm. The mere existence of the (coded) algorithm as a mathematical object is enough to satisfy the needs of the Platonist. For the Platonist, calculability is, as we already said, 'extensional', and pertains to the function as a mathematical *object*, but in no way as to how, if at all, that object is epistemically given to mathematical agents.[4] A constructivist mathematician, on the other hand, will insist on the algorithm being *known* to define a function, and hence, at least in principle, possible to execute. As so often in conflicts between Platonists and constructivists we have an instance of Fichte's (1797) dispute regarding epistemic priority between '*Dogmatists*' (that is, Realists or Platonists) and *Idealists*. The Platonist realist reduces the rightness of the epistemic act of knowledge to an object in the ontology, whereas for the constructivist the act is *sui generis* and yields the object of knowledge. Here, Platonist calculability, a modal notion, which is a matter of *agency*, is reduced to a mathematical property of an object, independently of how it is presented.

Constructivity, or constructiveness, is the property that matches the adjective *constructive* and my title noun *construction*. What is its range of significance? Or, in other words, to what can *constructive* be properly applied?[5] Mathematicians, positions, treatments, demonstrations, theorems, theories, and so on, are

---

These 'schemata of Turing's' need not be the rules for building Turing-machines. In the present context of giving notations for large natural numbers they could equally well, or perhaps even better, refer to 'the first rather general scheme for definition of number theoretic functions by transfinite recursion,' which, according to Robin Gandy, was given by Turing in his dissertation, cf. Feferman (1988, p. 141, footnote 24). Furthermore, Von Neumann mentioned Turing to Ulam several times 'concerning mechanical ways to develop formal mathematical systems.' Since the works of Gödel and Gentzen the addition of schemes for proof by transfinite induction or for defining functions by means of transfinite recursion are well-known ways for obtaining stronger mathematical systems.

Robin Gandy was not at Bletchley Park during the war. However, in the early 1950s he was working on Gentzen's First Consistency Proof, and would have had occasion to think about ordinal induction and recursion, cf. Kreisel (1955, footnote 9 at p. 38). His supervisor and friend Turing owned an offprint of Gentzen's paper; he might well have told Gandy about the general scheme of recursion from his Princeton dissertation and of the Bletchley challenge.

[4]Concerning Platonist extensionality, see, for instance, Rogers (1967, p. 9), or Enderton (2010, p. 5). The replacement of calculability by (Platonist) existence of a coded algorithm is an instance of the Platonist strategy of subsuming agency under (Platonist) existence of suitable objects.

[5]I confine myself to mathematical examples and shall disregard such uses as 'Neither his position, nor his conduct in the debate, could be called constructive', where 'constructive' seems to mean *apt* or *purposeful*.

sometimes called constructive and so are various mathematical objects; even *logic* has been called constructive. On the other hand, a mathematician—while doing his mathematics—would not usually say of the objects of study—for example, numbers, sets, and functions—that they are constructive. Indeed, one would not ordinarily call mathematical objects, theorems, or theories constructive. A function occurring in a constructive treatment would, in that context, just be called a function. One would single it out as *constructive* only when making a (meta-theoretical) contrast or comparison with other treatments, where also non-constructive functions may occur. Thus, for instance, in constructive and non-constructive mathematical theories alike, a function $+$ is found that is defined by the usual recursion equations. However, whether it is the *same* function occurs in both cases is a moot question. Michael Dummett, in his famous argument in favour of constructive, or (as he put it) 'intuitionistic', logic presupposes that there is a shared part between classical and constructive mathematics, and our $+$ function would belong to it.[6] However, is it a function in constructive mathematics or in classical mathematics? From what neutral position could the comparison be carried out? Does the idea of an over-arching, all-encompassing framework to which constructive and non-constructive denizens alike belong actually make sense? This is the Fichtean problem once again. Rather than considering the single function $+$ in two different frameworks (or in an all-encompassing, but possibly incoherent, framework of constructive and non-constructive entities alike) it might be better to consider two different functions $+_{\text{con}}$ and $+_{\text{class}}$ that belong to the respective frameworks. For the Platonist the $+_{\text{class}}$ function is a set of ordered pairs that is given by (or in?) the ontology of mathematics, *independently of its recursion equations*, and that ontology, rather than definitions and demonstrations, decides matters of right and wrong involving the $+_{\text{class}}$ function. For our Platonist, it is a mathematical truth, *but not a definition*, that the recursion equations hold for $+_{\text{class}}$. Such a Platonist *Ontological Descriptivism* is not shared in constructivist mathematics, where the $+_{\text{con}}$ function is introduced, or defined, by the recursion equations in question.[7] Prior to its definition being given, we simply have no $+_{\text{con}}$ function to reason about, even though it would of course have been (logically) possible to give the definition earlier than it was actually given. However, as already noted the point is indeed moot whether there is a neutral framework extending both the classical and the constructive ones and in which the functions $+_{\text{con}}$ and $+_{\text{class}}$ can be compared. Who would carry out such a comparison? The classical mathematician? Or the constructive mathematician? Neither would have access to the object of the other. But then who? Brouwer, who was an Ontological Descriptivist, but with respect to an *idealist* ontology, was aware of similar intricacies already in his inaugural lecture, where he compares classical and intuitionist readings of Cantor's diagonal demonstration that we cannot have a

[6] Dummett (1975, especially at p. 231).

[7] By Ontological Descriptivism I understand the philosophical position that the criteria of rightness for meaning, truth, and knowledge are ontologically obtained. It is spelled out a bit more in Sundholm and Van Atten (2008) and Sundholm (2013).

surjective mapping from the natural numbers **N** to the set of functions from $\mathbf{N} \to \mathbf{N}$.[8] This means that the picture of 'Bishop constructivism', which for present purposes we may take to be given by the meaning-theoretically based CTT, as the 'neutral kernel' of all kinds of constructivism, is not right. In particular, the equation

$$\text{Intuitionism} = \text{Bishop constructivism} + \text{Choice sequences} + \text{Continuity}$$

is thoroughly misleading.[9]

What, then, characterizes the notion of constructive when used in mathematics? W.W. Tait has put the matter very well:

> Let me say straight off that there are two distinct ideas: one is construction and the other is computation. These have been confused in recent history, but really are distinct. 'Constructive' means that the only witnesses of existential propositions one admits are ones that can be constructed, where of course this implies some background rules of construction. From the construction of an object, a means of computing it (in cases in which this idea makes sense) may or may not be found. In the context of arithmetic and analysis, constructive does imply computable (and this may indeed be the motivation for some to proceed constructively), but this is a theorem; it is not built into the notion of construction.[10]

With regard to formal treatments this should not be understood as a reference only to existentially quantified *propositions* of the form $(\exists x : D)B$. As is well known its proof-objects are of the form $\langle a, b \rangle$. In Martin-Löf's Constructive Type Theory (CTT) Tait's point is doubly taken care of by the use of existential quantifications via the $(\exists I)$ rule, that is, the existential quantifier introduction rule, as well as the form of judgement

$$\alpha \text{ exists}$$

where α is presupposed to be a type. The assertion condition for the judgement α **exists** is given by the rules of inference:

$$\frac{a \in \alpha}{\alpha \text{ exists.}}$$

Thus existence, 'instantiation', of a general concept ('type') may be claimed only when it has been instantiated.

Constructivism in mathematics holds that existence claims have to be supported by suitable means of instantiation: one is not entitled to assert that something exist without possession of a procedure that at least in principle produces an

[8]See Brouwer (1912). Brouwer's position thus is an epistemic realism based on an Ontological Descriptivism with respect to an idealist ontology. Bishop Berkeley is another philosopher with a similar pair of positions. My own meaning theoretical stance is to the contrary an epistemic idealism, whereas my ontology is thoroughly Platonist; however, I do not found my semantic and epistemic norms of rightness therein.

[9]Such equations are deployed by Beeson (1981, p. 148).

[10]Tait (2006, p. 213). The quoted text was italicized in its entirety by Tait.

example of what is said to exist. It is a rather natural position and with a venerable pedigree. Ancient precursors might include Proclus and Speusippos, and modern versions begin with Kronecker, and continue with Brouwer and Bishop. A constructive function, then, is a function as used in constructive mathematics, that is, mathematics that establishes existence claims by means of explicit instantiation. However, a distinction is called for here. First, there is the issue of a concrete function being constructive. Consider the number-theoretic function $! : \mathbf{N} \to \mathbf{N}$, that is, the 'factorial' that is well known to all students having taken an introductory course in statistics, and defined by the recursion equations

$$0! =_{\text{df}} 1$$
$$(n+1)! =_{\text{df}} n! \times (n+1).$$

It is clearly a calculable function: for every number x there is a number y such $x! = y$, and, in the familiar way, the recursion equations yield a procedure for calculating the value that is said to exist. Similar observations apply to all elements of *function sets* in Martin-Löf's Constructive Type Theory (CTT), that is, sets $A \to B$, where A and B are sets (predicatively generated from below).[11] Such an element $c : A \to B$ is a constructive function, since by the meaning-explanation for the set-builder $\to$, it will have to be evaluable to canonical form $\lambda x.b$, where $b : B$, given that $x : A$, and $\text{ap}(c, a)$, that is, the application of c to an argument $a : A$, is *definitionally equal* to $\text{ap}(\lambda x.b, a)$. But by the Π-equality rule—a kind of 'β-conversion'—$\text{ap}(\lambda x.b, a) =_{\text{df}} b[a/x] : B$.

As already required by Pascal, the notion of definitional equality $a =_{\text{df}} b$ admits effective replacement of *definiendum* a by its *definiens* b, whenever it occurs in an evaluation context.[12] Furthermore, the sets A and B are explained in terms of

[11]The dependent function-sets are formed with the use of the set-theoretic dependent-product operator Π, though for the sake of simplicity, I here confine myself to the $\to$ case where the set B does not depend upon the element of $x : A$. The elements of function-sets are known as *courses-of-value* or *graphs*, and require an application function ap in order to yield their values.

[12]Pascal gave his canon of definition in the manuscript *De l'Esprit géométrique*, but it became known through the *Port-Royal Logic*, Arnauld and Nicole (1662, Part IV, Ch. III). The notion of *definitional* (or *criterial*) identity $a = b : A$, with respect to the elements of the set A, must be clearly kept apart from the propositional function $I(A, x, y) : prop(x : A, y : A)$. From the judgement

$$a = b : A,$$

where the judgements $a : A$ and $b : A$ have to be already known as its presuppositions, we may, of course infer the judgement

$$I(A, a, b) \text{ true},$$

but not vice versa. The positions to the left of the judgemental colon are opaque in the terminology of Quine; there the terms carry '*conceptual* supposition' (*suppositio simplex* in the medieval terminology) and do not stand for their referents. To the right of the judgemental colon the positions are transparent and the matching supposition is 'referential' (*suppositio personalis*).

how their '*canonical* elements' may be put together out of parts. For any set the precise forms that its canonical elements have to take are known from its meaning explanation in terms of the constructors for the set in question. The definitional equalities tell us how to proceed towards the value, and the distinction between canonical and non-canonical elements allows us to determine when that value has finally been reached. In the apt terminology of Curry, that value is an *ultimate definiens*.[13]

Martin-Löf's CTT is the only smooth formal system for dealing constructively with functions.[14] In classical mathematics, where full Comprehension as well as classical logic are both freely available, functions can be dealt with as functional relations, and conversely sets and relations can be dealt with using characteristic functions. Thus, there it is largely a matter of convenience and taste what option to choose. The development in terms of predicates runs smoothly, and is much more familiar than the corresponding development in terms of functions, whence there are no substantial investigations to be found of higher-order function-theoretic logic. On the constructive side, the systems for intuitionistic analysis offered by Kleene (1965), Howard-Kreisel (1966), Kreisel-Troelstra (1970), and Scarpellini (1971), are quite cumbersome in comparison to second-order predicate logic, or to the great elegance of the fast classical development of analysis in Zermelo-Fraenkel set theory. In particular, apart from CTT, we have no readily accessible Natural Deduction system for function quantification.[15] In view of the contemporary fame and virtual omnipresence of Natural Deduction this is quite surprising.

How, and *where*, would one construct a 'non-constructive' function, that is, a function that could not be admitted in a constructivist framework? One way would make use of as yet *undecided* mathematical questions:

$$(\#)\qquad f(k) =_{\mathrm{df}} \begin{cases} 1 & \text{if the Riemann Hypothesis is true} \\ 0 & \text{if the Riemann Hypothesis is false} \end{cases}$$

---

[13]Curry and Feys (1958, p. 43).

[14]Together, of course, with other similar systems for typed lambda-calculus with dependent types. The richness of CTT allows one to distinguish three notions among what are commonly called functions in informal mathematics, to wit

(i) Frege-Euler functions, that is, dependent objects of lowest type, for instance $x + 5 : \mathbf{N}$ provided that $x : \mathbf{N}$, where substitution takes care of application;
(ii) Dedekind mappings, that is independent objects of higher type, for instance $(x).x + 5 : \mathbf{N} \to \mathbf{N}$, where application is a primitive notion;
(iii) Graphs, or '*courses-of-value*', for instance $\lambda(\mathbf{N}, \mathbf{N}, (x).x + 5) : (\Pi x : \mathbf{N})\mathbf{N}$, where application is effected by means of a separate application function $\mathrm{ap}(x, y)$ taking a course-of-value c and a suitable argument a into the value $\mathrm{ap}(c, a)$.

[15]The 1976 Basel dissertation by Robert Haberthür has remained a singular point here. I am indebted to Leon Geerdink for help in locating a copy of Haberthür (1976). See also Haberthür (1978).

and, with full generality, for any proposition A, making use of the type Bool of truth values:

$$(\#\#) \quad f_A(k) =_{\mathrm{df}} \begin{cases} \mathbf{t} & \text{if A is true;} \\ \mathbf{f} & \text{if A is false.} \end{cases}$$

In the absence of a constructive demonstration for

$$(A \vee \neg A) \text{ true}$$

we have no way to compute values of the function f_A.

Another way would be to introduce certain non-mathematical, *haphazard, empirical*, or *human* aspects in the definition, for instance (I am writing in the year AD 2013),

$$g(k) =_{\mathrm{df}} \begin{cases} 1 & \text{if } k < 2014 \text{ or the first child born in the year } k \text{ is a girl;} \\ 0 & \text{if } k > 2013 \text{ and the first child born in the year } k \text{ is a boy.} \end{cases}$$

These alleged definitions both yield 'functions' f and g whose values cannot be evaluated. For each particular number k, the first one is calculated by resolving the Riemann Hypothesis, whereas the second one may be calculated by waiting long enough, in some cases very long indeed. Note, however, that if human fertility were to fall drastically, the status of g as a function comes under pressure *also classically*. The presupposition of the definite description *the first child born in the year k* is after all a controversial one: when $k > 2013$, we do not *know* that a child will be born in year k.

Furthermore, a constructivist mathematician rejects purported definitions that proceed via a separation of undecided cases.[16] According to him, f is not a function, since it is not clear that f has values. Accordingly one is not entitled today to make the judgement

$$f : \mathbf{N} \to \mathbf{N},$$

since, according to its meaning explanation, inferences of the kind

$$\frac{k : \mathbf{N}}{f(k) : \mathbf{N}}$$

[16]Already Kronecker insisted that definitions should be, in the words of his pupil Jules (Julius) Molk (1885, p. 8) 'algebraic and not merely logical'. Molk also stressed that definitions by undecided separation of cases are inadmissible in his remarkable anticipation of Brouwer's 1908 criticism of the Law of Excluded Middle in (1904, '§10. *Point de vue de Kronecker*', at p. 160). Molk's scoop is dealt with at some length in the introduction, written jointly with Mark van Atten, to our novel translation of Brouwer 1908. Already Largeault (1993, p. 81) noted these little-known points about Molk, but did not stress their importance.

might be valid, but *we do not know*. Indeed, how does one evaluate $f(14)$? A classical mathematician, on the other hand, will hold that f *is* well defined, and that accordingly g has definite properties; thus, for instance, he will hold that $f(14)$ is a definite natural number, whence it is either odd or even, on the strength of the even constructively known theorem that $(\forall x : \mathbf{N})(x$ is odd or$x$ is even) is a true proposition. He will even hold that f is a primitive recursive function, *even though he is not able to compute any of its values*.[17] I would say that f has not been properly defined, whereas the classical mathematician says that according to its 'definition' f is either constant 1 or constant 0, whence in either case f is primitive recursive, and so f is primitive recursive. Let it be clear, furthermore, that not all definitions by means of a separation of cases are illegitimate from a constructive point of view. For instance, when one already knows that the proposition $A \vee B$ is true, it is correct to define a function from $\mathbf{N}$ to $\mathbf{N}$ by

$$g_1(k) =_{\mathrm{df}} \begin{cases} 1 & \text{if } A \text{ is true;} \\ 0 & \text{if } B \text{ is true.} \end{cases}$$

Similarly, when $(\forall x : \mathbf{N})(A(x) \vee B(x))$ is known to be true, the definition

$$g_2(k) =_{\mathrm{df}} \begin{cases} h_1(k) & \text{if } A(k) \text{ is true;} \\ h_2(k) & \text{if } B(k) \text{ is true} \end{cases}$$

is correct, etc.[18] The classical mathematician, who gives definitions via undecided separation of cases, while using locutions such as '$=_{\mathrm{df}}$ a when A is true and is $=_{\mathrm{df}}$ b *otherwise*', will of course claim that the required definitional presupposition, namely that the proposition $A \vee \neg A$ is true, is known as an instance of the classical Law of Excluded Middle.[19]

---

[17]An example of the kind g was famously used by Hartley Rogers in his canonical textbook on recursion theory (1967, §1.2, p. 9).

[18]In Martin-Löf's (1984) CTT, this is nothing but an application of the $\vee$-elimination rule.

[19]Undecided separation of cases is sometimes rejected also in other than mathematical contexts. Thus we do not accept—at least the fiscal authorities do not accept—the following as a definition of a human taxpayer:

$$\text{PM} =_{\mathrm{df}} \begin{cases} \text{Tony Blair} & \text{if the Riemann Hypothesis is true;} \\ \text{Gordon Brown} & \text{if the Riemann Hypothesis is false.} \end{cases}$$

PM does not pay taxes and is not accepted as an individual. Classical mathematicians, though, are readily prepared to proceed similarly when defining functions or sets.

When I first visited Holland in 1979, Henk Barendregt told me of the following nice example from the game of chess. It concerns a famous problem in 'retrograde analysis' by Langstaff from *Chess Amateur* 1922, *White to move and mate in two* (found on line here: http://en.wikipedia.org/wiki/Retrograde_analysis). If Black is not allowed to castle, that is, if either King or Rook has already moved, then 1. Kf5-e6 allows mate on the next move, but if Black is allowed to castle,

From a classical point of view, the existence of non-recursive functions is readily established by means of simple cardinality considerations. Primitive recursive functions are inductively defined by means of certain schemata. Thus, basic functions are in the class outright, and we have closure under composition as well as primitive recursion. These definitional schemata are without further presuppositions. One obtains a definition of the general recursive functions, under the guise of μ-recursive functions, by adding the definitional schema of *unbounded minimalization*: when ψ is a binary function, define a unary function ϕ by

$$\phi(x) =_{\text{df}} \mu y(\psi(x, y) = 0).$$[20]

In contradistinction to the schemata for primitive recursive functions, in order to guarantee the required totality of the function ϕ, applications of the *unbounded* μ-operator, does require a *presupposition*, namely the truth of $(\forall x : \mathbf{N})(\exists y : \mathbf{N})(\psi(x, y) = 0)$, that *must be known* prior to giving the definition in question.[21]

Just as in the primitive recursive case definitions via schemata give rise to an obvious indexing of the class of *recursive* functions. The *indexing* of primitive recursive functions is even primitive recursive, but, by an obvious diagonalization, its *universal function* is not primitive recursive. It can, however, be defined using unbounded minimalization, whence it is general recursive.[22] As is well known, there is no recursive universal function for the general recursive functions. In fact, not even the indexing will be recursive, or even effective, since the presupposition of totality that is required for applications of minimalization is Π^0_2, which takes us

then his last move was not with either King or Rook, and must accordingly be g7-g5, in which case 1. h5xg6 *en passant* allows mate on the next move. Tim Krabbé, the Dutch expert of the outré corners of chess, characterised the situation as follows (1985, p. 50–51): 'If White attempts one solution Black has a defence which shows the other would have worked. Or to put it differently again: it is perfectly true g7-g5 and Rh7-h8 cannot *both* be Black's last move, but White (or the solver) has no way of determining which one *was*. Whoever feels giddy should now consult his local syllogism breaker.' Krabbé is right that something out of the ordinary is going on in logic here. A constructivist would classify this as a typical LEM failure and reject the problem as ill posed.

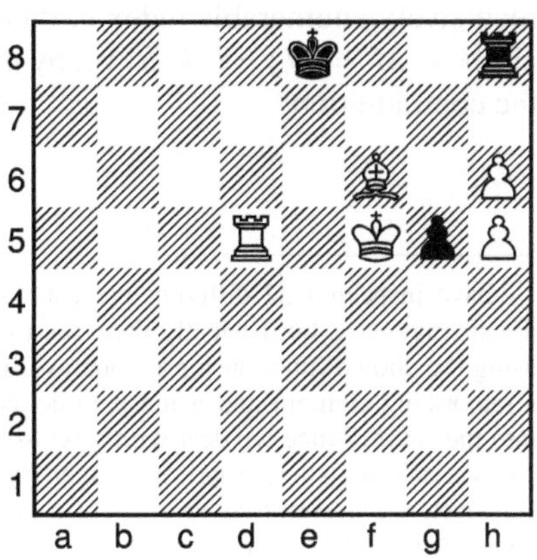

[20]For ease of exposition I display only the simplest form of the schema; each function may, of course, depend on a vector $x_1, \ldots, x_k$ of further variables.

[21]The need for the *known* truth of the presupposition here constitutes another argument against classical mathematical practice: unless known, mere truth of the totality presupposition does not entitle one to introduce the minimalization in question as a function.

[22]Kleene's indexing of the primitive recursive schemata of definition, and the diagonalization in question, can be found in his (1958).

straight out of the recursively decidable. Nevertheless, in view of this, albeit non-decidable, indexing by means of natural numbers, a classical mathematician will claim that the (general) recursive functions are countable, whereas the set of (all) functions from **N** to **N** is uncountable, by the original Cantorian diagonalization.

The definition of functions by means of undecided separation of cases is a major source of non-constructiveness in classical mathematics. Thus, the logical inferences in the proof of the Bolzano-Weierstrass theorem are constructive, and the non-constructive character of the result is due to an undecided separation of cases in the definition of the chain of nested closed intervals, each containing an infinitude of points from the given closed and bounded infinite set ('if the left half of the closed interval I_n contains infinitely many points from the set, choose it for I_{n+1}; otherwise choose the right half').

Zorn's Lemma provides another nice example of this phenomenon. Its various proofs make use of choice functions provided by Zermelo's Axiom of Choice; at some stage in the demonstration an undecided question is asked with respect to the choice function, and this renders the non-constructive character of the theorem manifest.[23]

The technique of using undecided separation of cases readily yields an explicit classical example of a non-recursive function. We use Kleene's recursively enumerable set K that is defined using his T-predicate:

$$K =_{\text{df}} \{n : \exists y T(n, n, y) \text{ is true}\}.^{24}$$

It is a fact, also constructively, that the set $M =_{\text{df}} \{n : \forall y \neg T(n, n, y) \text{ is true}\}$, that is, the classical complement of K, is *not* recursively enumerable. For if M has recursively enumerable index e, then, for all $k \in \mathbf{N}$, $k \in M \Leftrightarrow \exists y T(e, k, y)$, but also $k \in M \Leftrightarrow \forall y \neg T(k, k, y)$, by the definition of M. The choice of e for k gives the contradiction

---

[23]I have inspected 32 different presentations of Zorn's lemma and its demonstration. The fine details may vary, but basically there are two proofs. The easy one proceeds by ordinal recursion, using the choice function for as long as it is possible to find larger elements in the set m, to pick a majorizing element and using it to define an injection from an initial segment of the ordinals into the set m. Since we deal with a *set* m, it is not equinumerous with the set-theoretic universe, whence at some stage it will no longer be possible to find yet another majorizing element in m. However, the question as to whether a larger element can still be found within the set m is an undecided separation of cases. The other kind of proof is modelled after Zermelo's (1908) second, Dedekind-inspired, 'algebraic' demonstration of the well-ordering theorem and also here one finds definitions by means of undecided separations of cases.

[24]Kleene's T-predicates (for all number of arguments), and other associated functions and predicates, are primitive recursive (1952, § 57). In fact, they are even Kalmár-elementary. However, analogously to the above considerations regarding $+$ and its recursion equations, the significance of their Kalmár-elementary status depends on whether they are taken from a classical or a constructivist point of view: is it a mere factual property that happens to hold in the ontology or an essential means of introducing them?

$$\exists y T(e,e,y) \Leftrightarrow \forall y \neg T(e,e,y).$$

Now define the *characteristic function* of K

$$\chi_K(k) =_{\text{df}} \begin{cases} 1 & \text{if } k \in K; \\ 0 & \text{if } k \notin K. \end{cases}$$

Then $(\exists n : \mathbf{N})\chi((n) = 0) \Leftrightarrow k \notin K$. But, if the function χ_K is recursive, also the quantified matrix $\chi_K(n) = 0$ is recursive, and hence, in that case, K has a recursively enumerable complement, since the above equivalence gives a Σ_1^0 form. Hence the function χ_K is not recursive. From a constructive point of view, owing to the separation of cases, the definition is rejected and χ_K is not accepted as a function.

Church's Thesis is a crucial topic in the theory of computable functions and relates general recursiveness and computability:

> Every computable function is general recursive.

The converse direction is considered non-controversial. In fact, it is held to be more or less trivially correct, since its recursion equations provide an algorithm for the calculation of a given general recursive function. However, this might be too quick. When the demonstration that a function is general recursive is itself non-constructive, it may well be that the function will not be computable. As already noted above *a non-constructive demonstration that an algorithm (or code of algorithm) exists does not ensure calculability.* For instance, the function f above that is classically defined by means of an undecided question regarding the Riemann Hypothesis makes this clear: as we already saw the value of $f(14)$ cannot be calculated, whether by classical or constructive means. Hence, the classically defined function f, which is considered recursive by the classical mathematician, is not calculable. The sensitivity of calculability to the kind of demonstration offered for the existence of the algorithm was noted already by Alonzo Church when he introduced the Thesis:

> The reader may object that this algorithm cannot be held to provide an effective calculation of the required particular value of F^i unless the proof is constructive that the required equation ... will ultimately be found. But if so this merely means that he should take the existential quantifier which appears in our definition of a set of recursion equations in a constructive sense. What the criterion of constructiveness shall be is left to the reader.[25]

As noted by Kreisel (1987, p. 509) disagrees at this point; the matter was not left to the reader, but

> originally CT was intended and understood in the sense of ... effectiveness for the ideal mathematician ...

---

[25]Church (1936, p. 351, footnote 10). That calculability of a recursive function may depend essentially on the character of the demonstration that it is (primitive) recursive was lucidly stressed by Arend Heyting in a series of writings, for instance (1954, pp. 81–82), (1958b, p. 106), (1961, p. 187), (1962, p. 196), and (1969, p. 4).

Church's clarifying remark in the letter to Yannis Moschovakis in the appendix contradicts this:

> 'Church's Thesis' was originally proposed ... in the context of classical mathematics, with no thought of relating it to intuitionism or other form of constructivism, *either* in the sense of requiring a constructive proof or constructive interpretation of the existence proposition that was in question, *or* in the sense of using the thesis as a means of characterizing the (or a) notion of constructive proof.[26]

Furthermore, in the letter Church is also explicit on the importance of these issues for Kleene's realizability as a constructive interpretation of constructivism:

> Kleene later proposed to use the thesis in order to make a classical study of the notion of intuitionistic proof. This was done without altering the thesis by reconstructing its existential quantifier intuitionistically; and this led to objections from Heyting, who of course held that the existential quantifier must be so reconstructed.[27]

Against this background it is natural to ask:

> Is every function used in constructive mathematics recursive?

Great care is needed when answering this question. Heyting, who thought long and hard also about this and related questions in the 1950s, observed that the notion of a general recursive function cannot replace the primitive notion of a function in constructivism.[28] Every use of the minimalization operator to define a novel function demands that the presupposition be *known* to obtain. Since what has to be known is the truth of a Π_2^0 $\forall\exists$ proposition, we need its proof-object, but proof-objects for such propositions essentially involve functions. Hence, defining a general recursive function by minimalization presupposes the notion of a (constructive) function. Accordingly, it is only from a classical point of view that the notion of a (general) recursive function can serve as an explication of constructive functions. However, as we already saw above, also here one needs to have a *primitive* notion of constructive demonstration in order to ensure that recursive functions are computable.

Since algorithms can be coded arithmetically much of the above can, *mutatis mutandis*, be carried out also inside various formal systems for intuitionistic and constructive mathematics. For instance, one uses an analogue of $(**)$ above as a 'formal Church's Thesis'

$$(***) \quad (\forall\alpha : \mathbf{N} \to \mathbf{N})(\exists e : \mathbf{N})(\forall k : \mathbf{N})(\exists m : \mathbf{N})(T(e, k, n) \& U(m) =_{\mathbf{N}} \mathrm{ap}(\alpha, k)).$$

---

[26] Mendelson (1990, p. 226) stresses the classical origin of Church's Thesis: 'Thus, we are adopting a completely classical, nonintuitionistic stance'.

[27] Heyting voices a complaint of the kind alluded to by Church in (1954, pp. 81–82).

[28] Heyting made also this point on a number of occasions, including the pellucid (1958a). As observed by Thierry Coquand in his paper at the Joint Session, also Skolem saw that within constructivism the primitive notion of a function cannot be replaced by that of a general recursive function. Rózsa Péter (1959) is a third author stressing that the notion of a recursive function cannot constructively replace the notion of a constructive function.

In arithmetic, where function variables are lacking, one uses instead

$$(\forall x : \mathbf{N})(\exists y : \mathbf{N})A(x, y) \supset (\exists e : \mathbf{N})(\forall k : \mathbf{N})(\exists n : \mathbf{N})(T(e, k, n) \& A(k, U(n)).$$

Here the intensional Axiom of Choice (that is a derivable theorem in CTT), when applied to (***), yields a functional $\Psi : (\mathbf{N} \to \mathbf{N}) \to \mathbf{N}$, such that

(•) $(\forall \alpha : \mathbf{N} \to \mathbf{N})(\forall k : \mathbf{N})(\exists n : \mathbf{N})(T((\Psi(\alpha)), k, n) \& U(n) =_{\mathbf{N}} \mathrm{ap}(\alpha, k))$

Accordingly, when $I(\mathbf{N}, \Psi(\alpha), \Psi(\beta))$ true, from (•), we also have

(••) $(\forall x : \mathbf{N}) I(\mathbf{N}, \mathrm{ap}(\alpha, x), \mathrm{ap}(\beta, x))$ true.

This extensional equality, however, does not allow us to infer the conclusion that $I(N \to \mathbf{N}, \alpha, \beta)$ true. However, reasoning within classical set theory, the inference

$$\frac{\Psi(\alpha) =_{\mathbf{N}} \Psi(\beta) \text{ true} \Rightarrow (\forall x : \mathbf{N}) I(\mathbf{N}, \mathrm{ap}(\alpha, x), \mathrm{ap}(\beta, x)) \text{ true}}{\Psi(\alpha) =_{\mathbf{N}} \Psi(\beta) \text{ true} \Rightarrow \alpha =_{\mathbf{N}} \beta \text{ true}}$$

does hold, whence the functional Ψ is an *injection* of $\mathbf{N} \to \mathbf{N}$ into $\mathbf{N}$. Classically, via the Schroeder-Bernstein Theorem, this yields a bijection of **N** *onto* $\mathbf{N} \to \mathbf{N}$, and we have a contradiction with Cantor's Diagonal Theorem. So we have yet another demonstration that (Formal) Church's Thesis and classical reasoning are incompatible.[29] Kreisel's mammoth—'Saaty'—survey article (1965, §2, pp. 119–149) initiated a sometimes-intense debate on the foundations of intuitionistic analysis. William Howard and William Tait gave decisive contributions that originated in their work for the (unpublished) *Stanford Report* from 1963. At the third international congress for Logic, Methodology and the Philosophy of Science at Amsterdam, 1967, a rich symposium was devoted to 'Foundations of Mathematical Theories', and intuitionistic analysis figured prominently among these, with Kleene, Kreisel, Myhill, Tait, and Troelstra among the symposiasts.[30] For present purposes, the controversy regarding the so-called *Theory of the Creative Subject*, and the related Kripke's Schema, with their effects on Church's Thesis, is especially relevant to my topic here. In his lecture (1967) at the London 1965 conference on the Philosophy of Mathematics, Georg Kreisel gave an axiomatic formalization of Brouwer's view of the 'creating' (a Brouwerian term and preferable to *creative*), or 'thinking', subject. This formalization was not just put forward for meta-theoretic study, but was also intended to have foundational content. John Myhill was a tireless explorer of its possibilities, and it was speedily discovered that Kreisel's theory (or equivalently Kripke's schema) refuted (Formal) Church's Thesis. The debate culminated at the Buffalo 1968 meeting on *Intuitionism and Proof Theory*, with contributions by Hull, Kreisel, Myhill (1970), Van Rootselaar, Troelstra, and Vesley (1970). Troelstra's Buffalo lectures (1969) gave a broad overview, and, building on those, so did

[29]Beeson (1985, p. 90) presents a nice result of Troelstra's that Extensionality and Formal Church's Thesis are incompatible over type theory.

[30]See its Proceedings that were edited by Van Rootselaar and Staal (1968, Ch. 2, pp. 121–223).

Dummett (1977, Ch. 6.3), while Dirk van Dalen returned to the topic and added new twists.[31] Mark van Atten's (2004, Ch. 5) is probably the deepest and most thorough philosophical discussion of the Theory of the Creating Subject in print today.

That theory took off when Kreisel (1967) added a novel connective

$$\Sigma \vdash_m A$$

to the language of intuitionistic analysis, where $m : \mathbf{N}$ and Σ is the creating (or 'thinking') subject, also known as the *idealized mathematician*. The Σ does not serve as a variable here, but can perhaps best be thought of as being part of the (extended) turnstile $\Sigma \vdash$. The intended reading of '$\Sigma \vdash_m A$' was: *the (thinking) subject has evidence for asserting A at time m*. Troelstra (1969, §16, p. 95) changed this into 'the creative subject has evidence for A at stage m'. He also suggested a reading in terms of proofs: 'the creative subject has a proof of A at stage m.'[32]

Three axioms that we owe to Kripke and Kreisel (1967) for $\Sigma \vdash_m A$ have since become standard:

(CS1) For any proposition A, $\Sigma \vdash_x A$ is a decidable propositional function of A, that is, $(\forall x : \mathbf{N})(\Sigma \vdash_x A \vee \neg\Sigma \vdash_x A)$;
(CS2) $(\forall x : \mathbf{N})(\forall y : \mathbf{N})(\Sigma \vdash_x A \supset \Sigma \vdash_{x+y} A)$;
(CS3) $(\exists x : \mathbf{N})(\Sigma \vdash_x A) \Leftrightarrow A$.

Various weaker versions of CS3 were also considered, for instance, the contraposition of the controversial half of CS3, that is, $A \supset (\exists x : \mathbf{N})(\Sigma \vdash_x A)$ became known as the *Axiom of Christian Charity*:

(CC) $\neg(\exists x : \mathbf{N})(\Sigma \vdash_x A) \supset \neg A$.[33]

Kreisel's informal reading of the $\Sigma \vdash$ connective appears ill chosen. English alone uses 'evidence' in the familiar Anglo-Saxon *legal sense* of evidence *for* something. However, even in English, at least according to the *OED*, the first meaning of

[31] Van Dalen (1978, 1982a,b), and Troelstra and Van Dalen (1988, Ch. 4, §9).

[32] Van Atten (2008) also prefers to work in terms of proofs and uses 'The ideal mathematician has at time n obtained a proof of proposition p'. In his book (2004, p. 64) Van Atten, with an insignificant typographical change, uses 'the subject has experienced A at m', presumably in order to accommodate the Brouwerian view of truth (2004, p. 18): 'To experience a truth is to experience that a certain construction has succeeded.' This formulation might not be entirely felicitous: juxtaposing these passages, there seems to be no difference between experiencing a proposition and experiencing the truth of a proposition. In his joint paper with Van Dalen (2002, p. 522) he uses 'the creating subject experiences A (has full evidence for A) at time m', which appears to *identify experiencing A* with *having full evidence for A*. (For how to deal with evidence see further down the main text, around footnote mark 33.) The difference between the 'evidence' of earlier writers and the '*full* evidence' of Van Dalen and Van Atten is intended to rule out readings that admit insufficient evidence. Furthermore, the notion of proposition that is at issue here is not clear to me. It will not be the BHK one. Instead Van Atten's A would appear to contain assertoric force and be demonstrated.

[33] The responsibility for this 'ethico-theological' terminology is credited to Kreisel, though to the best of my knowledge it was first published by Myhill (1967, p. 296).

evidence is the *quality that pertains to what is evident*. Anglophone epistemologists attempt to save themselves by using 'self-evidence'. However, since *everything* that is evident, and not just what is *self*-evident, has the quality of evidence this move does not work. The evidence of a judgment made (demonstrated theorem) can also be discursive and obtained in several steps by means of a chain of immediate evidences, be they axiomatic or immediate inferences. Hence, evidence of, and not for, is the main epistemological notion to consider.[34] In the Anglophone legalistic sense one may well have evidence (that is, evidence *for* something) that is insufficient. In the case of mathematics, though, evidence in this 'legalistic' sense is clearly not enough, but a proper demonstration, or 'proof', is called for.

Kreisel (1967, p. 179) raises the issue of interpretation and notes that

> very little of the 'thinking subject' is used. ... Instead of writing $\Sigma \vdash_n A$, I could write $\Sigma_n \vdash A$ and read it as: the n-th proof establishes A. ... [T]he essential point would not be the individual subject, but the idea of proofs arranged in an ω-order.

Other readings were canvassed, for instance by Van Rootselaar (1970), who used $P(\sigma, n, A)$, instead of Kreisel's $\Sigma \vdash_m A$, but now with σ as a *variable* for 'creating subjects' and allowing iterated occurrences of P inside the proposition A.[35] Van Dalen and Van Atten in recent writings prefer to use the 'modal' *notation* $\Box_m$ in place of Kreisel's $\Sigma \vdash_m A$.

Van Rootselaar cast Kreisel's CS3 as

(3) $\exists\sigma(\exists m : \mathbf{N})P(\sigma, m, A) \Leftrightarrow A.$

and regards Kreisel's CC as a formal rendering of 'a well-known argument Brouwer's:

(4) *If the creative subject has evidence that he will never assert A,*
then he has evidence to assert $\neg A$'

and formalizes this using his P connective:

$$(CC^{\mathrm{vR}}) \quad P(\sigma, m, (\forall n : \mathbf{N})\neg P(\sigma, n, A)) \supset P(\sigma, m, \neg A).^{36}$$

[34]I deal with the ambiguities of *evidence* in my (2011).

[35]In the review (1967, p. 248), Kreisel allows for the quantification over creating subjects. His notation is still the same

$$\Sigma \vdash_n A$$

but now with Σ as a *variable* over subjects, and the reading: 'at time n the subject has evidence to assert A'. In (1970) he casts this as 'A has been proved by the nth stage'. In (1971) Kreisel elaborates on the alternative formulation in terms of an ω-ordering of demonstrations ('proofs') and notes that Kreisel (1970, §4) 'considers the schema KS which is inconsistent with C[hurch's] T[hesis]. (The schema KS was derived by Kripke from Brouwer's assertions about the thinking subject, or better from the postulate of an ω-ordering of levels of proofs.)'.

[36]Van Rootselaar (1970, p. 189), with altered fonts and logical symbolism. It does not seem unlikely that van Rootselaar had seen Kreisel (1967).

Van Rootselaar's treatment is an improvement, I think, on the Kreisel-Myhill framework, but neither is convincing. Both connectives—Kreisel's $\Sigma \vdash_x A$ as well as Van Rootselaar's extended $P(\sigma, n, A)$—stand in need of proper meaning explanations. From their use in the various axioms it is clear that both connectives have to be propositional in character, since otherwise they could not be used with quantifiers and connectives to form the propositions that occur in these axioms. However, neither Kreisel, or Van Rootselaar, nor, to the best of my knowledge, anyone else, has offered a suggestion towards a 'BHK' account for what canonical proof-objects might be for propositions built by means of such 'Creating Subject' connectives, and it is by no means obvious how to provide one. Furthermore, the informal explanations that were offered by Kreisel and others seem too vague to serve as a basis for an alternative notion of proposition that would serve as a *non-standard* interpretation of the constructive logical vocabulary. Accordingly, the Theory of the Creating Subject suffers from a considerable meaning-explanatory deficit at this point. Will the standard connectives and quantifiers have to be re-interpreted in order to accommodate the novel connective in either its Kreisel or Van Rootselaar version, or can they retain their standard BHK explanations? We simply do not know.

The question is also moot whether the Theory of the Creating Subject does not beg Geach's Frege point. Existence of a proof at stage m would not just give the truth of a proposition(al content), but actually entail knowledge that the proposition is true.[37] For instance, an ordinary natural deduction assumption that A is true (towards obtaining that B is true on condition that A is true) cannot then be made, since an assumption that there is a proof of A amounts to an assumption that A is known. The proofs available to the Creating Subject ('idealized mathematician') are too committing from an epistemic point of view, so to speak.[38]

Furthermore, with his use of the *variable* σ in the P connective, Van Rootselaar avails himself of object-level quantification over 'creative subjects'. In the absence of further explanation, one must assume that this is the standard constructive quantification, whence these creating subjects have to form a proper 'basic domain' D of quantification in the BHK sense. How, then, in that case, are the canonical elements of the domain of creating subjects generated from parts? At present we have no idea of how to answer this, to my mind, rather pressing question. The 'axioms' of the theory CS simply are not made evident by the meaning explanations.[39]

[37]Geach (1965).

[38]The same worry might apply also to a Brouwerian use of the implication $\supset$. For Brouwer, truth is truth known, 'experienced' truth, and accordingly an assumption that proposition A is true amounts to an assumption that A is known to be true. Van Atten (2009) contains a promising attempt at rescuing Brouwer's position by reading implication $A \supset B$ as essentially involving only relations between the proof-conditions, but not the actual existence of proofs, for the component propositions A and B.

[39]Here I part company with Van Dalen and Van Atten (2002, p. 522), who claim such evidence.

Already these meaning-theoretical worries are enough to impugn the theory from a constructive point of view. Equally worrisome, however, is that the 'axioms' proposed for the Creating Subject appear to be, not just non-evident, or lacking in justification, but incorrect. The controversial direction of CS3 is simply wrong to my mind. It does not hold as an implication, or equivalently as a consequence, but at most as a rule of proof, going from a known judgement to a novel judgement that gets known in this inference. The point is this: when an implicational proposition $A \supset B$ is asserted as an axiom, it allows one to pass also from *assumptions* that A is true to a conclusion that B is true.[40]

However, from an *assumption* that the proposition A is true we may *not* conclude that the Creating Subject, by carrying out, or otherwise obtaining, a suitable demonstration at a certain stage, gets to know that A is true. When the proposition A, whose truth is hypothetically assumed (as in natural deduction when one arrives at the dependent truth of B, under the assumption that the proposition A is true), is in fact false, of course, at no stage can the Creating Subject come up with a demonstration that A is true. There are no such demonstrations to be found. An inference, on the other hand, from *knowledge* that A is true, to knowledge that, at some stage, the Creating Subject knows that A is true, is not vitiated by the same error. Knowledge that A is true has to be based on possession of a proof(-object) a for A, and the Creating Subject might reflect on when it acquired that proof. Verification of the truth of an implication, or the holding of a consequence, be they logical or not, involves the construction of certain function(-object)s that transform proofs of antecedents into proofs of consequent, *irrespective of whether there actually are any proofs of* A; only a relation between the respective proof-conditions is involved, but not their being actually fulfilled. In the simplest case when x is assumed to be a proof of A, b is a proof of B, under that same assumption. Validation of an inference, on the other hand, involves assumptions not just that propositions are (hypothetically) true, but that judgements are (hypothetically) known.

The holding of consequence demands the construction of dependent proof-objects from *hypothetical*, assumed proof(-object)s. The validation of inferences, on the other hand, proceeds from hypothetical assumptions of *actual* proof-objects.

---

[40]Similarly it justifies the (open) consequence ('conditional')

if A is true, then B is true

that is, as the hypothetical judgement

B is true, on condition (hypothesis, assumption ...) that A is true,

which we may symbolize with a Gentzen sequent arrow

A true $\Rightarrow$ B true.

Constructively an open consequence (hypothetical judgment, Gentzen sequent) is verified by a hypothetical ('dependent') proof-object $b : B(x : A)$.

When we divest the calculi of content, and consider them purely formally, these points on implication and consequence may be cast also purely syntactically in terms of *formal derivations*. The rule

$$\frac{\Gamma \vdash A}{\Gamma \vdash (\exists m : \mathbf{N})(\Box_m A)}$$

which is equivalent to the controversial direction of the CS3 axiom, is not valid.[41] Removing the assumptions Γ, just as in the case of the rule of Necessitation in modal logic, leads to a rule that perhaps can be justified:

$$\frac{\vdash A}{\vdash (\exists m : \mathbf{N})(\Box_m A).}$$

When the Creating Subject knows that A is true he does so on the basis of a possessed proof-object, and this proof-object is obtained at a certain stage that can be determined by introspection.[42] However, in an interpreted calculus, with content, the worries about the propositionality of $\Box_m$ of course remain.

Analogous observations can be made on the so-called 'Fitch's paradox', where the crucial 'verifiability' axiom $A \supset \Diamond KA$ simply is not valid as a constructive axiom. First, also here there is the meaning-theoretical issue about the propositionality of K and $\Box$. In the absence of a BHK account of these notions the Fitchean anti-anti-realist's attempted refutation of anti-realism may be safely ignored by the anti-realist. Again, fulfillability of the proof-condition of A is not needed for verifying the truth of an implication $A \supset B$, but only the establishment of a certain relation between the proof-conditions of antecedent proposition A and consequent proposition B. When A is only *assumed* to be true, why should it then be possible to know that A is true? For all we know A might actually be false. Yet this is granted by the Fitchean 'axiom' in question. The conditional

$$A \text{ is true } \supset \text{ it is possible to know that } A \text{ is true}$$

does not hold in general; the assumption of a hypothetical proof-object $x : Proof(A)$ does not warrant the inference that A can be known. In natural-deduction terms we have here the difference between taking ϕ as an assumption wff towards deriving ψ, in order to obtain $\phi \supset \psi$ by means of an $\supset$I application that cancels the assumption ϕ, and assuming that we possess a closed derivation of ϕ. An assumption

[41]In order to prevent misunderstanding I here prefer to use the Van Dalen-Van Atten 'modal' notation for the Creating Subject connective rather than the Kreisel-Myhill-Dummett turnstile notation.

[42]Kreisel, Myhill, and others have given a considerable body of derivations using the axiomatic formulation of CS, conveniently spelled out in Troelstra (1969) and Dummett (1977); I have not checked which of these derivations still go through using only the corresponding rule of proof.

that x is a hypothetical proof of A does not guarantee that A true can be known, that is, that an actual proof-object a of A can be found.

On the other hand, the *rule of inference*

$$\frac{\vdash\ A \text{ is true}}{\vdash\ \text{it is possible to know that } A \text{ is true,}}$$

is correct without assumptions to the left of the turnstile, or, in natural deduction terms, when the premise that A is true is known outright under no assumptions. Validity of inference does not consist in mere preservation of propositional truth from antecedent propositions to consequent proposition; rather it is the possibility to know the conclusion judgement given that the premise judgements are known that is at issue.

The Theory of Creative Subject was suggested in order to formalize Brouwerian 'strong counter-examples'. Saul Kripke discovered that its use may be avoided here, using instead Kripke's Schema.[43] The 'negative' formulation of Kripke's Schema I prefer to use here was,[44] in essence, introduced by Van Dalen (1982b), but I make the role of the two truth-values **t** and **f** explicit via the type **Bool**, in order to bring out the analogy with undecided separation of cases:

$$\text{(KS)}\quad (\exists\alpha \in \mathbf{N} \to \mathbf{Bool})[A \Leftrightarrow (\exists x : \mathbf{N})(\alpha(x) \neq_{\mathbf{Bool}} \mathbf{f})]$$

where $x =_{\mathbf{Bool}} y =_{\mathrm{df}} I(\mathbf{Bool}, x, y)$, that is, propositional identity with respect to the set **Bool**.

The function that is asserted to exist in (KS) is analogous to the non-constructive function f_A at (##) that tests the proposition A. Van Dalen (1982b) pointed out that KS is correct from a classical point of view:

> Kripke's Schema looks a bit like a trivial comprehension principle. The Comprehension Principle postulates a characteristic function that tests whether an element has the property $A(x)$. Kripke's Schema postulates a function that tests whether A holds. From a classical standpoint that is wholly trivial.[45]

---

[43]The letter from Moschovakis to Church printed in the appendix makes clear that Kripke's reasoning was known in 1968, at the time of the Buffalo conference on *Intuitionism and Proof Theory*. The schema is clearly presaged in Kripke (1965).

[44]By 'negative' I understand using $\neq_{\mathbf{Bool}}$ **f** rather than $=_{\mathbf{Bool}}$ **t** when formulating the principle.

[45]Van Dalen (1982b, p. 174, my translation):

> Kripkes Schema sieht ein bißchen aus wie ein triviales Komprehensionsprinzip. Das Komprehensionsprinzip postuliert eine charakteristische Funktion, die prüft, ob ein Element die Eigenschaft $A(x)$ hat. Kripkes Schema postuliert eine Funktion, die prüft, ob A gilt. Klassisch ist das natürlich völlig trivial.

I am indebted to Mark van Atten, who drew my attention to this passage when informed of the classical derivation of Kripke's Schema given above. Troelstra and Van Dalen (1988, Ch 4, §9) use second-level quantification, not just over functions in **Bool** → **N**, or in **N** → **N**, but also over sets ('*species*') instead and treat of Kripke's schema in terms of the quantifier combination

The classical definition (##) of f_A is indeed trivial. However, the matching derivation of Kripke's Schema, using a proof $?(A)$ of $(A \vee \neg A)$, is a bit more instructive, whence I give one here.

We define the function $\alpha : \mathbf{N} \rightarrow \mathbf{Bool}$ by means of an undecided separation of cases. For $k : \mathbf{N}$, put

$$(*) \quad \alpha(k) =_{\mathrm{df}} \begin{cases} \mathbf{t} & \text{if } A \text{ is true;} \\ \mathbf{f} & \text{if } A \text{ is false.} \end{cases}$$

(1) Assume A true. Then by $(**)$
(2) $[\alpha(0) =_{\mathbf{Bool}} \mathbf{t}]$ true, but
(3) $[\mathbf{t} \neq \mathbf{f}]$ true, and so by the rules for $=_{\mathbf{Bool}}$
(4) $[\alpha(0) \neq_{\mathbf{Bool}} \mathbf{f}]$ true, and whence by $\exists$-introduction
(5) $(\exists x : \mathbf{N})[\alpha(x) \neq_{\mathbf{Bool}} \mathbf{f}]$ true.

For the other direction,

(6) Assume $(\exists x : \mathbf{N})[\alpha(x) \neq_{\mathbf{Bool}} \mathbf{f}]$ and
(7) Assume $x : \mathbf{N}$ such that $[\alpha(x) \neq_{\mathbf{Bool}} \mathbf{f}]$ true.
(8) Assume A false. Then, by $(**)$
(9) $[\alpha(x) =_{\mathbf{Bool}} \mathbf{f}]$,

whence we have a contradiction between (7) and (9), so

(10) $\perp$ true.

But then, using $\perp_c$ on (8) and (10),

(11) A true. By $\exists$-elimination on (6), (7) and (11), discharging (6) we have
(12) A true.

So by $\Leftrightarrow$I on (1) and (5), (6) and (12) we have

(13) $[A \Leftrightarrow (\exists x : \mathbf{N})(\alpha(x) \neq_{\mathbf{Bool}} \mathbf{f})]$, and by $\exists$-introduction
(14) $(\exists \alpha \in \mathbf{N} \rightarrow \mathbf{Bool})[A \Leftrightarrow (\exists x : \mathbf{N})(\alpha(x) \neq_{\mathbf{Bool}} \mathbf{f})]$, that is KS.

In order to complete the demonstration we must deal with the method of *Definition by undecided separation of cases* as applied in $(**)$.

Let A, B be given propositions. First we define a function

$$\mathrm{sep} : A \vee B \rightarrow \mathbf{Bool}$$

$$(\forall X : prop)(\exists \alpha : \mathbf{N} \rightarrow \mathbf{N})(\forall n : \mathbf{N})$$

(where I have made quantificational domains explicit). From a constructive point of view this introduces yet another, and this time superfluous, complication into the Theory of the Creating Subject, namely that of impredicative quantificational domains, whence I prefer the earlier treatments in terms of functions, especially in the closed (BKP) form.

by the equations

$$\text{sep}(i(a)) = \mathbf{t} : \mathbf{Bool}\ (a : A)$$
$$\text{sep}(j(b)) = \mathbf{f} : \mathbf{Bool}\ (b : B)$$

Then put

$$g(x, z) = \text{sep}(z) : \mathbf{Bool}\ (x : \mathbf{N}, z : A \vee B)$$

Finally, we *postulate* a non-constructive LEM-proof

$$\frac{A : \mathbf{prop}}{?(A) : \text{Proof}(A \vee \neg A).}$$

Putting together all the components we obtain the function for Kripke's Schema:

$$\alpha_A(x) = g(x, ?(A)) : \mathbf{Bool}\ (x : \mathbf{N}).$$

By starting the above demonstration with a propositional variable X instead of a given proposition, we obtain a closed verification object c_0 for Kripke's Schema, but now formulated as a closed principle (that I think may be appropriately named after Brouwer and Kripke):

(BKP) $(X : \mathbf{prop})\text{Proof}((\exists\alpha \in \mathbf{N} \to \mathbf{Bool})[X \Leftrightarrow (\exists x : \mathbf{N})(\alpha(x) \neq_{\mathbf{Bool}} \mathbf{f})])$.

This ('functional') generality with respect to propositions is not quantificational, since the domain of propositions does not form a set, whence it is not amenable to quantification. Thus, the c_0 above is a *verification*-object, but not a *proof*-object, since BKP is not a proposition. The construction of this particular c_0 makes use of classical logic in the form of the assumed verification-object for LEM

$$? : (X : \mathbf{prop})\text{Proof}(X \vee \neg X).$$

A demonstration of Kripke's Schema, however, does not need the *full* power of classical logic, as given by the function ? that yields a proof-object $?(A)$ of $A \vee \neg A$, when applied to a proposition A. The more modest resources of CS 1–3 will suffice. The CS1 decidability of $\Sigma \vdash_x A$ allows us to replace the undecided separation of cases in (##) by a decidable one: for $m : \mathbf{N}$, define

$$\alpha(m) =_{\text{df}} \begin{cases} \mathbf{t} & \text{if } \Sigma \vdash_m A \text{ is true;} \\ \mathbf{f} & \text{otherwise,} \end{cases}$$

where the legitimacy of the definition is guaranteed by CS1, and KS follows readily.[46]

Already in Myhill (1967, pp. 296–297) it was pointed out that KS is incompatible with (formal) Church's Thesis, and Troelstra (1969, pp. 98, 100) gives more results with such a recursion-theoretic flavour. In view of the decidability-axiom CS1 that allows one to use a decidable separation of cases in place of the undecided separation of cases that is used to give classical characterizing functions, I find it hardly surprising that KS allows for non-recursive functions. Both Myhill, and following him Troelstra, note that these results are not really recursion-theoretic in character. Instead they are analogues to results concerning the impossibility of enumerations within various quantifier prefixes. Such phenomena are familiar already from classical predicate logic. Thus, for instance, the logically valid formula

$$\neg\exists y \forall x (R(x,y) \Leftrightarrow \neg R(x,x))$$

embodies reasoning used in various paradoxes, e.g., those of Russell and Grelling.

Saul Kripke gave an actual example of a function, using the Theory of the Creating Subject for its definition, thereby establishing the incompatibility of CS with Church's Thesis in a stronger way.[47] Above we saw how to define the characteristic function of Kleene's non-recursive but recursively enumerable set K, using undecided separation of cases. Exactly the same kind of construction is now used for Kreisel's connective $\Sigma \vdash_m A$, which by stipulation is decidable, so the separation of cases is no longer an undecidable one.

$$G(m,n) =_{\text{df}} \begin{cases} 1 & \text{if not } \Sigma \vdash_m n \notin K; \\ 0 & \text{if } \Sigma \vdash_m n \notin K. \end{cases}$$

Then $n \notin K \Leftrightarrow (\exists m : \mathbf{N})[G(m,n) = 0]$.

From right to left, by the definition of G, when $G(m,n) = 0$, the Creating Subject has a proof of $n \notin K$ at stage m, and so $n \notin K$. From left to right, using the Brouwerian conception of truth, $n \notin K$ has been proved by the idealized mathematician, and so he has a proof of $n \notin K$. But this was obtained at some stage m and so for some m, $\Sigma \vdash_m n \notin K$, and so $(\exists m : \mathbf{N})[G(m,n) = 0]$. If the function G is recursive, the complement of K is recursively enumerable, and so K will be recursive, which is a contradiction. Accordingly the function G, which is held to be computable by the idealized mathematician, is not recursive. In view of my doubts both as to the meaningfulness and validity of the Kreisel-Kripke axioms, I am not

[46] As far as I know the precise proof-theoretic strength of Kripke's schema is undetermined.

[47] Kripke, to the best of my knowledge, never published his treatment, and perhaps it was not even written up. Van Dalen (1978, p. 40, footnote 3) gave a nice exposition that is unfortunately marred by garbled printing. Van Dalen's presentation was emended by Van Atten (2008), whose exposition I follow.

at all convinced by the interpretation of the Kripke result that makes a computable function out of G, and instead I prefer to view this result as a version of the classical theorem that there are non-recursive functions.

As already noted above, from its interaction with the BHK propositional connectives and quantifiers in the CS axioms it is clear that $\vdash_m$ has to be propositional.[48] Accordingly one might have expected a BHK account in the early writings, but none was forthcoming.[49] Can something be done about rendering the Brouwer-Kreisel connective $\Sigma \vdash_m A$ more palatable from a meaning-theoretical point of view? Dirk van Dalen (1978; 1982a) sketched model theoretic proofs that KS is conservative with respect to intuitionistic analysis and HA.[50] In his metamathematical constructions infinitely many novel constants β_A are used that serve as witnessing 'Kripke functions' for the existential quantifier in KS—one for each wff A. These 'Kripke functions' cannot be used at a contentful level, since that would require infinitely many semantic primitives.

At this point the reformulation of KS into BKP proves helpful. The assumption of a closed verification-object c for BKP allows us to obtain the required Kripke functions uniformly and contentfully. Let A be a proposition. We define

$$d =_{\mathrm{df}} c(A) : (\exists \alpha \in \mathbf{N} \to \mathbf{Bool})[A \Leftrightarrow (\exists x : \mathbf{N})(\mathrm{ap}(\alpha, x) \neq_{\mathbf{Bool}} \mathbf{f})],$$

whence

$$c_A =_{\mathrm{df}} p(c(A)) : \mathbf{N} \to \mathbf{Bool}$$

and

$$d_A =_{\mathrm{df}} q(c(A)) : A \Leftrightarrow (\exists x : \mathbf{N})(\mathrm{ap}(c_A, x) \neq_{\mathbf{Bool}} \mathbf{f}).$$

These witnessing Kripke functions c_A *are obtained uniformly* in the verification-object c of BKP and the chosen proposition A and make it possible to define the Kreisel-Myhill connective explicitly as a proposition in Martin-Löf's CTT. It must be stressed, though, that the meaningfulness is a relative one: given a verification-object for Kripke's Schema it is possible to define the connective. Whether such a verification-object can be found constructively is still an open question.

---

[48]Myhill refers explicitly and at length to BHK in (1967).

[49]This holds true also for quantification with respect to choice sequences; one would have had hopes for a uniform BHK account of quantification with respect to all quantificational domains and then an explanation of the notion of choice sequence such that, for instance, the various continuity principles can be read off by combining these explanations. Heyting's brief treatment (1959, §6 pp. 70–71) is the only one by a Founding Father known to me. The issue is noted in my (1983, pp. 163–164) and (1984, pp. xiii–xiv).

[50]Dummett's (1977, p. 353, formula (xxi*)) explains Van Dalen's construction, but also does not give details.

$$\frac{A \text{ is a proposition } m : \mathbf{N}}{(\Sigma \vdash_m A) \ \mathbf{prop}}$$

$$(\Sigma \vdash_m A = (\exists x \leq m : \mathbf{N})[\mathrm{ap}(c_A, x) \neq_{\mathbf{Bool}} \mathbf{f}]) \ \mathbf{prop}.$$

Here

(i) $x \leq y : \mathbf{prop}\ (x : \mathbf{N}, y : \mathbf{N})$, and the usual arithmetical properties of $\leq$ are demonstrable already in HA, but HA is embedded in CTT, so they are demonstrable also in CTT. In HA we define $x \leq y$ by $(\exists z : \mathbf{N})(z + x =_{\mathbf{N}} y)$, but in CTT with one universe we also have other options, for instance, using primitive recursion to define the propositional function $x \leq y$ outright.

(ii) When the quantifier matrix is decidable, that is, when we have a proof-object

$$\gamma : (\forall x : \mathbf{N})(A(x) \vee \neg A(x)),$$

also bounded $\exists$ and $\forall$ quantifiers are decidable. In particular, we can exhibit a proof

$$\eta : (\exists x \leq m : \mathbf{N})A \vee \neg(\exists x \leq m : \mathbf{N})A.$$ [51]

(iii) The propositional identity $\neg_{\mathbf{Bool}}$ with respect to the Boolean values **t** and **f**, that is, the propositional function

$$I(\mathbf{Bool}, x, y)\ (x : \mathbf{Bool}, y : \mathbf{Bool})$$

is decidable in CTT with one universe, since we have a proof-object

$$\tau : (\forall x : \mathbf{Bool})(\forall y : \mathbf{Bool})[x =_{\mathbf{Bool}} y \vee \neg x \neq_{\mathbf{Bool}} y].$$ [52]

[51] Kleene has given meticulous HA treatments of the relevant arithmetical requirements in (1952, §32, *150, p. 191) and Kleene–Vesley (1965, Remark 4.1, p. 15).

[52] In general, when P is a propositional function of the appropriate type, inferences of the kind

$$\frac{I(A, a, b) \text{ true}}{P(a) \leftrightarrow P(b) \text{ true}}$$

are valid. Accordingly, we define the function $h : \mathbf{Bool} \to U$, where U is the first universe, by **Bool** elimination:

$$h(\mathbf{t}) = \top : U$$

$$h(\mathbf{f}) = \bot : U.$$

We now demonstrate the axioms CS1–CS3, still under the epistemic assumption of a closed verification-object for BKS. Without it a definition of $\Sigma \vdash_m A$ might not even be possible.

CS1. The propositional function $B(x) = [\mathrm{ap}(c_A, x) \neq_{\mathbf{Bool}} \mathbf{f}]$ $(x : \mathbf{N})$ is decidable since $\neq_{\mathbf{Bool}}$ is decidable by (iii). Hence, by (ii), the Brouwer-Kreisel connective

$$\Sigma \vdash_m A =_{\mathrm{df}} (\exists x \leq m : \mathbf{N})[\mathrm{ap}(c_A, x) \neq_{\mathbf{Bool}} \mathbf{f}]$$

is decidable and CS1 is true.

CS2. Assume that $\Sigma \vdash_m A$ is true. By definition, that is, assume that

$$(\exists x \leq m : \mathbf{N})[\mathrm{ap}(c_A, x) \neq_{\mathbf{Bool}} \mathbf{f}] \text{ is true.}$$

Since standard properties of $\leq$ are demonstrable in HA, whence also in CTT,

$$(\forall m : \mathbf{N})(\forall n : \mathbf{N})[m \leq m + n] \text{ true}$$

Hence, from the assumption

$$[\mathbf{t} =_{\mathbf{Bool}} \mathbf{f}] \text{ true}$$

we get that their respective values under the function h are propositionally equivalent, whence $(\top \leftrightarrow \bot)$ true, and so that $\bot$ true under the assumption. Hence $(\mathbf{t} \neq_{\mathbf{Bool}} \mathbf{f})$ true. The crucial Peano axiom

$$[\mathbf{s}(x) \neq_{\mathbf{N}} 0] \text{ true}$$

is demonstrated by the same technique, that is, defining a function from $\mathbf{N}$ to the first universe U that is $\bot$ on 0 and $\top$ on the numbers greater than 0.

In order to demonstrate the decidability of $=_{\mathbf{Bool}}$, assume $x : \mathbf{Bool}$ and $y : \mathbf{Bool}$. By the **Bool** elimination rule we readily demonstrate

$$\forall z : \mathbf{Bool}(z =_{\mathbf{Bool}} \mathbf{t} \vee z =_{\mathbf{Bool}} \mathbf{f}) \text{ true.}$$

But then $x =_{\mathbf{Bool}} \mathbf{t} \vee x =_{\mathbf{Bool}} \mathbf{f}$).

When (Case 1) $(x =_{\mathbf{Bool}} \mathbf{t})$ true, we have two cases, (1a) $(y =_{\mathbf{Bool}} \mathbf{t})$ true, and (1b) $(y =_{\mathbf{Bool}} \mathbf{f})$ true.

In (Case 1a), $(x =_{\mathbf{Bool}} \mathbf{t})$ true and $(y =_{\mathbf{Bool}} \mathbf{t})$ true, whence $(x =_{\mathbf{Bool}} y)$ true by $=_{\mathbf{Bool}}$ rules, and so, by $\vee$ elimination, $(x =_{\mathbf{Bool}} y \vee x \neq_{\mathbf{Bool}} y)$ true.

In (Case 1b), $(x =_{\mathbf{Bool}} \mathbf{t})$ true and $(y =_{\mathbf{Bool}} \mathbf{f})$ true. An assumption that $(x =_{\mathbf{Bool}} y)$ true by means of the rules for $=_{\mathbf{Bool}}$ yields $\mathbf{t} =_{\mathbf{Bool}} \mathbf{f}$, which gives a contradiction with the already demonstrated theorem that $(\mathbf{t} \neq_{\mathbf{Bool}} \mathbf{f})$ true. Hence $(x \neq_{\mathbf{Bool}} y)$ true, and by $\vee$ elimination, $(x =_{\mathbf{Bool}} y \vee x \neq_{\mathbf{Bool}} y)$ true also in this case.

When (Case 2) $(x =_{\mathbf{Bool}} \mathbf{f})$ true, there are yet again two cases.

In (Case 2a) $(x =_{\mathbf{Bool}} \mathbf{f})$ true and $(y =_{\mathbf{Bool}} \mathbf{t})$ true. As in (Case 1b) we obtain $(x \neq_{\mathbf{Bool}} y)$ true and so, by $\vee$ elimination, $(x =_{\mathbf{Bool}} y \vee x \neq_{\mathbf{Bool}} y)$ true. In (Case 2a) $(x =_{\mathbf{Bool}} \mathbf{f})$ true and $(y =_{\mathbf{Bool}} \mathbf{f})$ true, whence $(x =_{\mathbf{Bool}} y)$ true. By $\vee$ elimination we obtain $(x =_{\mathbf{Bool}} y \vee x \neq_{\mathbf{Bool}} y)$ true and two applications of $\forall$ introduction with respect to the assumptions $x : \mathbf{Bool}$ and $y : \mathbf{Bool}$ establish the required decidability of $\neq_{\mathbf{Bool}}$.

is demonstrable. Accordingly we can demonstrate the consequence

$$x \leq m \text{ true } \Rightarrow x \leq m + n \text{ true},$$

whence the truth of CS2 follows immediately.

CS3. By BKS, since A is a proposition, we get that

$$c(A) : (\exists\alpha \in \mathbf{N} \to \mathbf{Bool})[A \Leftrightarrow (\forall x : \mathbf{N})(\text{ap}(\alpha, x) \neq_{\mathbf{Bool}} \mathbf{f})],$$

and so

$$c_A = p(c(A)) : \mathbf{N} \to \mathbf{Bool}$$

and

$$d_A =_{\text{df}} q(c(A)) : A \Leftrightarrow (\forall x : \mathbf{N})(\text{ap}(c_A, x) \neq_{\mathbf{Bool}} \mathbf{f}).$$

Now, assume that A is true. Hence $(\exists m : \mathbf{N})(\text{ap}(c_A, m) \neq_{\mathbf{Bool}} \mathbf{f})$ true and so $(\exists m : \mathbf{N})(\exists x \leq m : \mathbf{N})[\text{ap}(c_A, x) \neq_{\mathbf{Bool}} \mathbf{f}]$, so $(\exists m : \mathbf{N})(\Sigma \vdash_m A)$ true.

For the other direction of CS3, assume that $(\exists m : \mathbf{N})(\Sigma \vdash_m A)$ true, that is, assume $(\exists m : \mathbf{N})(\exists x \leq m : \mathbf{N})[\text{ap}(c_A, x) \neq_{\mathbf{Bool}} \mathbf{f}]$ is true. Then $(\forall x : \mathbf{N})$ $[\text{ap}(c_A, x) \neq_{\mathbf{Bool}} \mathbf{f}]$ true, and by BKS, A is true.

With this reduction of the meaningfulness and main CS properties of the Kreisel-Myhill connective $\Sigma \vdash_m A$, a formal demonstration has been given that the Brouwer-Kripke Principle is 'sufficient for deriving most of the counterexamples of Brouwer': these counter-examples can be given in the CS, and so BKP suffices.[53] It will not have escaped the reader's attention that as a matter of fact a closed verification-object for BKP was obtained in its classical demonstration from the assumed closed verification-object ? for the Law of Excluded Middle above. Accordingly, via the reduction of the theory CS to the Brouwer-Kripke Principle, the formal Theory of the Creating Subject is *classically* valid, a somewhat surprising result since 'it represents the extreme consequences of intuitionistic subjectivism'.[54]

Acknowledgements I was an invited speaker at the Joint Session in Paris and spoke on the notion of function in constructive mathematics. However, problems of health in 2007–2008 unfortunately prevented me from submitting anything to its Proceedings, whence it was a pleasant surprise, as well as a challenging task gladly undertaken, when Michel Bourdeau and Shahid Rahman requested that I write an introductory essay for the Proceedings volume. Accordingly, in these *Afterthoughts* I return to some of the things I said in my Paris talk in 2006, as well as present later reflections, caused by rethinking some of the issues and rereading many of the original sources. The material on Brouwer's Creating Subject that is presented here was not part of my Paris lecture. It stems from research carried out during a Visiting Professorship at Lille, February to April 2012, and I am grateful to my host Shahid Rahman and his students for being such a keen audience. The

[53] Hull (1969, p. 241).

[54] The words are Troelstra's (1969, p. 107).

criticism of the Kreisel-Troelstra-Dummett account was first presented in a Lille seminar April 5, 2012, with, appropriately enough, several of the participants of the Paris Joint Session also present, to wit, my Paris hosts Van Atten, Bourdeau, and Fichot.

I am indebted to Yannis Moschovakis for his valuable comments in the Paris discussion, but above all for sharing with me his correspondence with Alonzo Church and granting permission to publish this important document. I am also indebted to Thierry Coquand for discussion of Heyting's views, as well as for providing me with a copy of the little known paper by Skolem that is referred to in his contribution to the present volume. Andrew Hodges and Tim Krabbé answered questions respectively on Alan Turing and on Langstaff. Michèle Friend gave helpful comments on the first, non-technical part. Furthermore, as always in recent years, I am indebted to Mark van Atten and Per Martin-Löf for stimulating conversations and valuable comments. In particular, Van Atten, by means of incisive questioning, forced me to revise my formulations on Leibniz and *calculemus*. He, as wel as Zoe McConaughey, helped greatly in producing the final draft of the manuscript. The IHPST, Paris, in the persons of Michel Bourdeau and Jean Fichot, offered generous hospitality on two occasions in May and November 2012.

Appendix

Correspondence between Yannis Moschovakis and Alonzo Church in 1968.

Professor Moschovakis reviewed four papers on Church's thesis for the JSL. The papers were 'Quelques Pseudo-Paradoxes de la "Calculabilité Effective"' by Jean Porte, 'An Argument Against the Plausibility of Church's Thesis' by Laszlo Kalmár (1959), 'Rekursivität und Konstruktivität' by Rózsa Péter and *On some recent criticism of Church's Thesis* by Elliott Mendelson (1963).

The review, which is highly relevant review to our topic, was printed in *The Journal of Symbolic Logic*, Vol. **33**:3 (Sep., 1968), pp. 471–472. Many of the topics addressed in Moschovakis' review and in the ensuing correspondence have been addressed in the main text of my paper, though I deliberately refrain from discussing Kalmár's non-plausibility argument against Church's Thesis that is conducted in terms of a purported counter-example. In his brief article Kalmár discusses or makes use of many themes that have been covered, for instance, the use of Kleene's set K, definition of functions by means of undecided separation of cases, and the nature of assumptions that something is known. He also makes use of an assumption analogous to Kreisel's reading of CS in terms of the existence of an 'enumeration' of demonstrations. A proper treatment of the complex dialectical structure of Kalmár's reasoning, together with a review of the not inconsiderable discussion that was provoked by his paper, would double the length of the present paper. Accordingly, in order not to try the Editorial patience beyond endurance, it shall be left for another occasion.

I. Holograph letter from Moschovakis to Church, submitting the review of the four papers.

March 15, 1968

Dear Prof. Church,

...

You might be interested in one of the mathematical points that originally caused me some difficulty in writing this up and which I still have not resolved entirely to my satisfaction. Kalmár's argument appears silly on first reading and may very well be silly through and through. It is rather close, however, to some of the arguments that Brouwer used in his later years concerning the so-called 'creating agent'. Kripke has extracted from these arguments the following principle, whose formal version is sometimes called Kripke's Schema:

> if P is any proposition, then there exists a (constructive) sequence $k_0, k_1, \dots$ of integers such that P is true if and only if some $k_i = 0$.

Apparently Brouwer's motivation is that the sequence $k_0, k_1, \dots$ can be 'constructed' (empirically and not by a mechanical procedure) by the 'creating agent' in his effort to find proofs for P; he sets $k_n = 0$ if at time n he has found a proof. This is very close to Kalmár's argument, except that Kalmár uses *classical reasoning* to obtain his 'paradoxical' conclusion. This is a very big difference, since the entire justification for Kripke's Schema (to the extent that it can be justified) lies in the inuitionistic interpretation of absurdity: P is absurd if no proof for P can be found (by any creative agent, in all time, presumably). In the end I decided that the explicit use of the classical ideas by Kalmár made his argument sufficiently different so I left the analogies with these ideas of Brouwer out of the review. But it is interesting to note that within the kind of intuitionistic mathematics that accepts Kripke's Schema, Kalmár's algorithm is acceptable. (Kripke's Schema has been used to violate formal versions of Church's thesis in systems with variables for 'constructive' sequences by Myhill and Kreisel, I believe.) I do not know if Kalmár knew of the Brouwer ideas and whether he was influenced.

Sincerely yours

Yannis N. Moschovakis.

II. Typed reply by Alonzo Church to the above letter.

April 15, 1968

Dear Professor Moschovakis,

This is just to thank you again for your review of Porte, Kalmár, et. al., and to make in partial reply to your letter the following historical remarks.

1. 'Church's Thesis' was originally proposed (Bulletin of the A.M.S. vol. 41(1935), pp. 332–333) in the context of classical mathematics, with no thought of relating it to intuitionism or other form of constructivism, *either* in the sense of requiring a constructive proof or constructive interpretation of the existence proposition that was in question, *or* in the sense of using the thesis as a means of characterizing the (or a) notion of constructive proof. Indeed in the historical context of the date it might well be said that the notions of *calculability* (the adjective 'effective' is superfluous) and *provability* had long existed in a pre-formal way in classical mathematics. But it can at most very doubtfully be said that a notion of constructive provability existed in *classical* mathematics. Kleene later proposed to use the thesis in order to make a classical study of the notion of intuitionistic proof. This was done without altering the thesis by reconstructing its existential quantifier intuitionistically; and this led to objections from Heyting, who of course held that the existential quantifier must be so reconstructed.

2. Though I have no definite information, I think it unlikely that Kalmár's argument was suggested or influenced by the ideas of Brouwer to which you refer. This is simply on the ground that Kalmár's mathematical publications have never shown any great concern with | intuitionism. And certainly his argument in the paper under review, as you report it, is the

very antithesis of intuitionism. That is, he assumes in effect that, for every x, either there is a proof of Ax (by some correct means) or there is a proof of the negation of Ax (by some correct means, in each case). And this is a special case of, and clearly kindred in spirit to, Hilbert's principle of the solvability of every mathematical problem, which Brouwer once so roundly denounced. I am not familiar with the Myhill-Kreisel paper to which you refer, but I suppose that their proof can hardly proceed in a way that parallels Kalmár's, or at least not unless they take Hilbert's principle or some special case of it as an axiom.

Very sincerely yours

Alonzo Church.

Added Note, April 2014

1. In the paper I follow Van Dalen's use of the negative formulation of Kripke's Schema. However, the 'positive' version of the principle

$$(X : \mathbf{prop})((\exists\alpha : \mathbf{N} \to \mathbf{Bool})[X \Leftrightarrow (\exists k : \mathbf{N})(\alpha(k) =_{\mathbf{Bool}} \mathbf{t})])$$

is more elegant and, in view of the decidability of $=_{\mathbf{Bool}}$, would work as well.
2. The function $P : (\mathbf{Bool})\mathbf{prop}$ that allows us to go from a boolean truth-value to the matching proposition.

$$P(\mathit{true}) = \top : \mathbf{prop}$$
$$P(\mathit{false}) = \bot : \mathbf{prop}.$$

It is well known that the Law of Excluded Middle

$$A \vee \neg A \; true \; (A : \mathbf{prop})$$

is equivalent to having a (unary) truth-value function $\beta : \mathbf{prop} \to \mathbf{Bool}$ such that

$$[P(\beta(A)) \leftrightarrow A] \; true.$$

Per Martin-Löf (email March 30, 2013) commented that the analysis of the Theory of the Creating Subject weakens this version of the Law of Excluded Middle by replacing the demand for the unary function β from **prop** to **Bool** to the demand for a binary function α, from **prop** *and* **N** to **Bool**, such that

$$[(\exists n : \mathbf{N}) P(\alpha(A, n)) \leftrightarrow A] \; true$$

instead. 'One could say', so Martin-Löf, 'that by an exact analysis of the mathematical content of the theory of the Creating Subject, you unravel it as a weakened form of the Law of Excluded Middle.'

3. Mark van Atten (January 2014) drew my attention to the fine analysis of the theory of the Creating Subject offered by Anne Troelstra (1980) in 'The interplay between logic and mathematics: intuitionism'. Troelstra notes (pp. 208–2099) that, with given function constants ψ_A, the Kripke equivalences $A \leftrightarrow \exists x (\psi_A x = 0)$ can be demonstrated from classical comprehension axioms, and that Kripke's Schema does not contradict classical logic. My principal contribution here is to provide such closed Kripke functions using the classical verification-object for Kripke's Principle.

References

Ackermann, W. (1928). Zum Hilbertschen Aufbau der reellen Zahlen. *Mathematische Annalen, 99*, 118–133.

Arnauld, A., & Nicole, P. (1662). *La Logique ou L'Art de Penser*. Paris: C. Savreux.

van Atten, M. (2004). *On brouwer*. Louisville: Wadsworth—Thomson Learning.

van Atten, M. (2008). The foundations of mathematics as a study of life: An effective but non-recursive function. *Progress in Theoretical Physics, 173*, 38–47.

van Atten, M. (2009). The hypothetical judgement in the history of intuitionistic logic. In C. Glymour, W. Wang, & D. Westerståhl (Eds.), *Logic, methodology, and philosophy of science XIII: Proceedings of the 2007 international congress*, Beijing (pp. 122–136). London: College Publications.

Beeson, M. (1981). Formalizing constructive mathematics: Why and how? In F. Richman, *Constructive mathematics* (Lecture notes in mathematics, Vol. 873, pp. 146–190). Berlin: Springer.

Beeson, M. (1985). *Foundations of constructive mathematics*. Berlin: Springer.

Brouwer, L. E. J. (1912). *Intuitionisme en Formalisme* (Inaugural lecture at the University of Amsterdam, October 14, 1912). Amsterdam: Clausen (Translated into English by Arnold Dresden as Intuitionism and formalism. *Bulletin of the American Mathematical Society, 20*, 81–96 (1913), reprinted in both editions of P. Benacerraf & H. Putnam (Eds.), *Philosophy of Mathematics*, Blackwell, Oxford, 1964, pp. 66–77, and Cambridge University Press, Cambridge, 1983, pp. 77–89.

Church, A. (1936). An unsolvable problem in elementary number theory. *American Journal of Mathematics, 58*, 345–363

Curry, H., & Feys, R. (1958). *Combinatory logic* (Vol. 1). Amsterdam: North-Holland.

van Dalen, D. (1978a). An interpretation of intuitionistic analysis. *Annals of Mathematical Logic, 13*, 1–43.

van Dalen, D. (1978b). *Filosofische grondslagen van de wiskunde*. Assen: Van Gorcum.

van Dalen, D. (1982a). The creative subject and heyting arithmetic. In *Universal algebra and applications* (Banach Center publications, Vol. 8, pp. 379–382). Warsaw: PWN Scientific Publishers.

van Dalen, D. (1982b). Braucht die konstruktive Mathematik Grundlagen? *Jahresbericht der Deutschen Mathematiker-Vereinigung, 84*, 57–78.

van Dalen, D., & van Atten, M. (2002). Intuitionism. In D. Jaquette (Ed.), *A companion to philosophical logic* (pp. 513–530). Oxford: Balckwell.

Dummett, M. (1975). The philosophical basis of intuitionistic logic. In M. Dummett, *Truth and other Enigmas* (pp. 215–247) London: Duckworth, 1978. Text of a lecture delivered at the Bristol *Proceedings of the logic Colloquium'73*, and originally published in the 1975 thereof.

Dummett, M. (1977). *Elements of intuitionism*. Oxford: Oxford University Press.

Enderton, H. (2010). *Computability theory: An introduction to recursion theory*. Burlington: Academic.

Feferman, S. (1988). Turing in the land of O(z). In R. Herken (Ed.), *The Universal Turing Machine* (pp. 113–147). Oxford/New York: Oxford University Press.
Fichte, J. G. (1797). Erste Einleitung in die Wissenschaftslehre. *Philosophisches Journal, V*, 1–47.
Geach, P. (1965). Assertion. *Philosophical Review, 74*(4), 449–465.
Haberthür, R. (1976). *Normalizationssätze für Intuitionistische Theorien mit Wahlfolgen*. Dissertation, Basel University.
Haberthür, R. (1978). Choice sequences and reduction processes. *Archiv für mathematische Logik, 19*, 31–49.
Heyting, A. (1954). Logique et intuitionnisme. In *Actes du 2e colloque internationale de logique mathématique,* Paris, 1952 (pp. 75–82). Paris-Lovain: Gauthier-Villars.
Heyting, A. (1958a). Blick von der intuitionistischen Warte. *Dialectica, 12*, 332–345.
Heyting, A. (1958b). Intuitionism in mathematics. In R. Klibansky (Ed.), *Philosophy in the mid-century. A survey* (pp. 101–115). Firenze: La Nuova Editrice.
Heyting, A. (1959). Some remarks on intuitionism. In A. Heyting (Ed.), *Constructivity in mathematics* (pp. 72–80). Amsterdam: North-Holland.
Heyting, A. (1961). Infinitistic methods from a finitist point of view. In *Infinitistic methods* (pp. 185–192). Oxford/Warsaw: Pergamon/PWN.
Heyting, A. (1962). After thirty years. In E. Nagel, P. Suppes, & A. Tarski (Eds.), *Logic, methodology, and philosophy of science, proceedings of 1960 international congress*, Stanford (pp. 194–197). Stanford University Press.
Heyting, A. (1969). Wat is berekenbaar? *Nieuw Archief voor Wiskunde (3), 17*, 1–7.
Hodges, A. (1983). *Alan turing: The Enigma*. London: Burnett Books.
Howard, W., & Kreisel, G. (1966). Transfinite induction and bar induction of types zero and one, and the role of continuity in intuitionistic analysis. *Journal of Symbolic Logic, 31*, 325–358.
Hull, R. (1969). Counterexamples in intuitionistic analysis using Kripke's schema. *Zeitschrift für mathematische Logik und Grundlagen der Mathematik, 15*, 241–246.
Kalmár, L. (1959). An argument against Church's thesis. In A. Heyting (Ed.), *Constructivity in mathematics* (pp. 72–80). Amsterdam: North-Holland.
Kleene, S. (1938). On notation for ordinal numbers. *Journal of Symbolic Logic, 3*, 150–155.
Kleene, S. (1952) *Introduction to metamathematics*. Amsterdam: North-Holland.
Kleene, S. (1958). Extension of an effectively generated class of functions by enumeration. *Colloquium Mathematicum, 6*, 67–78.
Kleene, S., & Vesley, R. (1965). *The foundations of intuitionistic mathematics*. Amsterdam: North-Holland.
Krabbé, T. (1985). *Chess curiosities*. London: George Allen and Unwin.
Kreisel, G. (1955). Models, translations, and interpretations. In Th. Skolem et al. (Eds.), *Mathematical interpretations of formal systems* (pp. 26–50). Amsterdam: North-Holland.
Kreisel, G. (1965). Mathematical logic. In T. Saaty (Ed.), *Lectures on modern mathematics* (pp. 95–195). New York: Wiley.
Kreisel, G. (1967a). Informal rigour and completeness proofs. In I. Lakatos (Ed.), *Problems in the philosophy of mathematics. Proceedings of the international colloquium in the philosophy of science*, London, 1965 (Vol. 1, pp. 138–186). Amsterdam: North-Holland.
Kreisel, G. (1967b). Review of Kleene and Vesley (1965). *Zentralblatt für Mathematik und ihre Grenzgebiete, 123*, 246–248.
Kreisel, G. (1970a). Church's thesis: A kind of reducibility axiom for constructive mathematics. In A. Kino, J. Myhill, & R. E. Vesley (Eds.), *Intuitionism and proof theory. Proceedings of the summer conference*, Buffalo, 1968 (pp. 121–150). Amsterdam: North-Holland.
Kreisel, G. (1970b). Review of Myhill (1967). *Zentralblatt für Mathematik und ihre Grenzgebiete, 187*, 263–265.
Kreisel, G. (1971). Review of Kreisel (1970). *Zentralblatt für Mathematik und ihre Grenzgebiete, 199*, 300–301.
Kreisel, G. (1987). Church's thesis and the ideal of informal rigour. *Notre Dame Journal of Formal Logic, 28*, 499–519.

Kreisel, G., & Troelstra, A. S. (1970). Formal systems for some branches of intuitionistic analysis. *Annals of Mathematical Logic, 1*, 229–387.

Kripke, S. A. (1965). Semantical analysis of intuitionistic logic I. In J. Crossley & M. Dummett (Eds.), *Formal systems and recursive functions* (pp. 92–130). Amsterdam: North-Holland.

Largeault, J. (1993). *Intuition et intuitionisme*. Paris: Vrin.

Martin-Löf, P. (1984). *Intuitionistic type theory*. Naples: Bibliopolis. Notes taken by Giovanni Sambin of lectures given a Padua, 1980.

Mendelson, E. (1963). On some recent criticisms of Church's thesis. *Notre Dame Journal of Formal Logic, 4*(3), 201–205.

Mendelson, E. (1990). Second thoughts about Church's thesis and mathematical proofs. *Journal of Philosophy, 88*(5), 225–233.

Molk, J. (1885). Sur une notion qui comprend celle de la divisibilité et sur la théorie de l'élimination. *Acta Mathematica*, **6**, 1–166.

Molk, J. (1904, 1907). Nombres irrationnels et notion de limite. Translated, expanded, and emended version of A. Pringsheim's German original, Chapitre 1–3. In J. Molk (Ed.), *Encyclopédie des Science Mathématiques Pure et Appliquée* (Vol. 1)—Arithmétique, *Fascicule 1* (August 10, 1904, pp. 133–160), *Fascicule 2* (May 20, 1907, pp. 161–208).

Myhill, J. (1967). Notes towards an axiomatization of intuitionistic analysis. *Logique et Analyse, 35*, 280–297.

Myhill, J. (1968). Formal systems of intuitionistic analysis I. In B. van Rootselaar & J. F. Staal (Eds.), *Logic, methodology and philosophy of science III. Proceedings of the third international conference for logic, methodology and philosophy of science*, Amsterdam, 1968 (pp. 161–178).

Myhill, J. (1970). Formal systems of intuitionistic analysis II: The theory of species. In A. Kino, J. Myhill, & R. E. Vesley (Eds.), *Intuitionism and proof theory. Proceeedings of the summer conference*, Buffalo, 1968 (pp. 151–162). Amsterdam: North-Holland.

Péter, R. (1959). Rekursivität und Konstruktivität. In A. Heyting (Ed.), *Constructivity in mathematics* (pp. 226–233). Amsterdam: North-Holland.

Rogers, H. (1967). *Theory of recursive functions and effective computability*. New York: McGraw-Hill.

van Rootselaar, B. (1970). On subjective mathematical assertions. In A. Kino, J. Myhill, & R. E. Vesley (Eds.), *Intuitionism and proof theory. Proceeedings of the summer conference*, Buffalo, 1968 (pp. 187–196). Amsterdam: North-Holland.

van Rootselaar, B., & Staal, J. F. (Eds.). (1968). *Logic, methodology and philosophy of science III. Proceedings of the third international conference for logic, methodology and philosophy of science*. Amsterdam: North-Holland.

Scarpellini, B. (1971). *Proof theory and intuitionistic systems* (Lecture notes in mathematics, Vol. 212). Berlin: Springer.

Sundholm, G. (1983). Constructions, proofs and the meaning of the logical constants. *Journal of Philosophical Logic, 12*, 151–172.

Sundholm, G. (1984). Brouwer's anticipation of the principle of charity. In *Proceedings of the Aristotelian Society, 85*(1984–1985), 263–276.

Sundholm, G. (2011). The vocabulary of epistemology, with observations on some surprising shortcomings of the English language. In A. Reboul (Ed.), *Philosophical papers dedicated to Kevin Mulligan*, Genève. www.philosophie.ch/kevin/festschrift/Sundholm-paper.pdf.

Sundholm, G. (2013). Demonstrations versus proofs, being an afterword to constructions, proofs, and the meaning of the logical constants. In M. S. van der Schaar (Ed.), *Judgement and the epistemic foundation of logic* (pp. 15–22). Dordrecht: Springer.

Sundholm, G., & van Atten, M. (2008). The proper explanation of intuitionistic logic: On Brouwer's proof of the Bar theorem. In M. van Atten, P. Boldini, M. Bourdeau, & G. Heinzmann (Eds.), *One hundred years of intuitionism (1907–2007). The Cerisy conference*. Basel: Birkhäuser.

Tait, W. (2006). Gödel's interpretation of intuitionism. *Philosophia Mathematica (III), 14*, 208–228.

Troelstra, A. (1969). *Principles of intuitionism* (Lecture notes in mathematics, Vol. 95). Berlin: Springer.

Troelstra, A. (1980). The interplay between logic and mathematics: Intuitionism. In E. Agazzi (Ed.), *Modern logic. A survey* (pp. 197–221). Dordrecht: D. Reidel.

Troelstra, A., & van Dalen, D. (1988). *Constructivism in mathematics* (Vol. I). Amsterdam: North-Holland.

Vesley, R. (1970). A palatable substitute for Kripke's schema. In A. Kino, J. Myhill, & R. E. Vesley (Eds.), *Intuitionism and proof theory. Proceedings of the summer conference*, Buffalo, 1968 (pp. 197–207). Amsterdam: North-Holland.

Zermelo, E. (1908). Neuer Beweis für die Möglichkeit einer Wohlordnung. *Mathematische Annalen, 65*, 107–128.

Zweig, S. (1979). *The right to Heresy*. In *Erasmus and the right to Heresy*. London: Souvenir Press (orig. edition Constable, 1936).

Chapter 2
The Developments of the Concept of Machine Computability from 1936 to the 1960s

Jean Mosconi

Abstract From the 1940s to the 1960s, despite the significant work done on recursive functions (properly) and later on the lambda-calculus, the theory of calculability was developed more and more as a theory of computation by an idealized machine, or in the form of a general theory of algorithms. I will only deal here with the former aspect, a development that stems from the concepts introduced in 1936 by Turing. I will try to show how Turing's ideas were gradually adopted, developed and modified. The Turing machine had an increasingly important role and was the object of systematic investigation. It was subsequently reworked to such an extent that a new model of machine was fashioned, the program and register machine. However, the initial model kept a significant place, and extensions of Turing's analysis led, toward the end of the century, to profound reflections about the notion of a constructive object and the general notion of an algorithm.

2.1 Introduction

Since 1936, several precise and coextensive concepts of computability, or effective calculability, have been in competition, and depending on the time and period some concepts have been studied and used more often than others. Robert I. Soare (1996) points out the following paradox: a large and increasing part of what was usually named "Recursion Function Theory" dealt in fact with computability by finite processes, and had as its basic technical notion the concept of computability by an idealized machine, which Turing had proposed in 1936. According to Soare (1996, 314), computability is, indeed, "the heart of the subject", which properly must be called "Computability Theory".

To a large extent, I agree with Soare's view of "Computability Theory", but my present purpose is neither critical nor normative, only historical. I shall try to show how, between the 1940s and mid 1960s, Turing's concept of machine

J. Mosconi (✉)
UMR 8560 IHPST – Institut d'Histoire et de Philosophie des Sciences et des Techniques, Université Paris 1 Panthéon-Sorbonne & CNRS & ENS, Paris, France
e-mail: jean.mosconi@outlook.fr

J. Dubucs, M. Bourdeau (eds.), *Constructivity and Computability in Historical and Philosophical Perspective*, Logic, Epistemology, and the Unity of Science 34,
DOI 10.1007/978-94-017-9217-2_2

computability was adopted, studied, developed and reworked. At first, Turing's model of computability was not widely followed, but it progressively gained importance and was subjected to a systematic investigation and reworking that led to a model of computation noticeably different from the initial one.

2.2 A Quiet Beginning (1936–1947)

2.2.1 A Philosophically Well Received, but Seldom Used Concept

The concept of computability by an idealized machine, namely a finite device operating in a deterministic and absolutely elementary way and storing information on an infinite tape, was introduced by Turing in 1936 in his paper "On Computable Numbers, with an Application to the Entscheidungsproblem". The enthusiastic philosophical reception of Turing's approach stands in stark contrast to the very limited attention given to it in print in the following decade. Among the various concepts proposed as a precise counterpart of effective calculability, the Turing machine was the most immediately adequate: as Church remarked in his review (Church 1937, 43), "it has the advantage of making the identification with effectiveness in the ordinary (not explicitly defined) sense evident immediately – i.e. without the necessity of proving preliminary theorems". It revealed itself to be the best intuitive support to Church's Thesis or, more exactly, to Turing's Thesis; this was already emphasized by Gödel shortly after Turing's paper was published (Gödel 193?, 168) and also later in the 1964 Postscriptum to his 1934 Lectures: "That this is the correct definition of mechanical computability was established beyond any doubt by Turing" (Gödel 1934, 369). For the concept of a Turing machine was introduced together with a rigorous analysis of the conditions imposed on any human operator carrying out a computation according to any algorithm. Hao Wang (1974), R. Gandy (1980; 1988; 2001) and especially W. Sieg (1994), Sieg and Byrnes (1996; 1999) gave us very illuminating commentaries on Turing's approach of calculability, which they themselves extended or transposed. Nothing like Turing's analysis was provided by Post to support his (technically very similar) concept of "finite combinatory process" (Post 1936).

However, when Post (1947, 2) resorted to "the theory of Turing machines" (a phrase seemingly not used before), what had been published earlier in the domain was more or less limited to: (a) Turing's original 1936 paper and the one that immediately followed it (Turing 1937), in which equivalence to λ-definability was proved; (b) Turing's 1939 paper that introduced the notion of an oracle machine and thus the related notion of relative computability (relative to some possibly non-recursive set), later investigated by Post in 1944; and (c), in the extended sense of "Turing machine", Post's short note of 1936.

Nonetheless, as early as 1943, Post considered that Turing machines should play an important role in the development of computability theory. Kleene (1989, 163) said that from his 1941 seminar he used a "considerably reworked" form of Turing machines, that became familiar to us in 1952 through Kleene's book.

2.2.2 *Early Revisions*

In his 1947 paper, Post emphasizes the usefulness of a Turing machine in studying Thue's problem, but he also stresses that, because of its awkwardness and defectiveness and also because its lack of generality, it cannot be used directly. Hence Post uses a revised version of it, which nonetheless keeps the original tape-machine structure of Turing (1936). In a long appendix to his paper, Post carries out a tight critical examination of Turing's machine concept. He shows that some technical details are incorrect and that the definition presents serious gaps or inconsistencies. For example, it is not explicitly stated that the "infinite tape" is only semi-infinite to the right and not two-way infinite. Also, according to Turing's general definition, a Turing machine starts on an empty tape, but this cannot be the case for the universal Turing machine. More importantly, Post remarks, Turing's convention about alternate squares (F-squares bearing non-erasable, final figures, 0's or 1's, E-squares bearing other, erasable symbols for scratch work) has unfortunate consequences, e.g. making some sets (of machine description numbers) non-recursively enumerable that would be recursively enumerable were it not for this convention. In a later paper (1950), Turing admits that Post's objections were justified on the whole.

In Turing's definition, the task that was given to the machine was writing a "computable real number", i.e. an infinite sequence of 0's or 1's separated by the E-squares. Of course, as Turing showed, we could adapt the machine so that it computes the value of a function f of natural numbers, but in an awkward way. Furthermore, since the machine's task would consist in writing down the infinite sequence $f(0), f(1), \dots$ of the values of the function, it could not be used to compute partial recursive functions (introduced by Kleene in 1938). This is what brought Kleene to conceive a Turing machine as working in successive stages: applied to a tape containing the argument, it computes, if possible, the value of the function, writes it next to the argument and stops; if it cannot compute, it runs on forever (Kleene 1981a). Gödel later emphasized that this notion of partial computable function is actually more adequate, as a characterization of mechanical computability, than that of total computable function, because the correct notion of computation need not presuppose that a computation terminates in every case (cf. Hao Wang 1974, II 3). Moreover, this definition avoids debates about the classical vs. constructivist interpretation of the existential quantifier involved in the definition of a total function.

2.3 The Rise of the 'Theory of Turing Machines' During the 1950s

After 1950, the increasing interest in the concept of a Turing machine takes various forms and is due to the several virtues this concept has.

2.3.1 The Convenience of the Concept of a Turing Machine for Expository Purposes

Turing's approach is gradually becoming recognized as being "intrinsically convincing" (Kleene 1981b, 49) by a more general audience. For example, in R. Péter's *Rekursive Funktionen* (1951), while the whole of the book deals with recursive functions (properly), the writer added a chapter on "berechenbare Funktionen" (i.e. functions computable by Turing machines) to show the adequacy of the concepts she introduced for intuitive calculability. In fact, Péter accepted a limited form of Turing's Thesis: functions that are mechanically computable by mathematical means available today are computable by Turing machines.

Hermes (1961) takes the concept of a Turing machine as its main thread in that it provides "the most natural and easy access" to the theory of calculability. Kleene's *Mathematical Logic* (1967) is another example. Hartley Rogers in his 1967 book (*Theory of Recursive Functions and Effective Computability*) deems it unnecessary to give any justification for such a choice: "Turing's characterization will be taken as basis in this book". Moreover, it is increasingly apparent that the Turing machine can provide the most natural and direct link between computability theory and actual computers.

2.3.2 The Mathematical Fruitfulness of the Concept of a Turing Machine

2.3.2.1 The Methodological Relevance of Turing's Unsolvability Proof

In his undecidability proof for first-order logic, Turing starts by constructing unsolvable problems concerning Turing machines: the halting problem, the problem of determining whether a given machine will ever print 0. He then shows that one can construct a first-order formula that describes the behavior of a Turing machine from a given initial configuration in the following sense: the formula is provable in first-order logic if and only if the machine eventually prints 0. Hence first-order provability is not decidable (by means of any Turing machine).

Bernays (1958) stressed that this proof gives a simple example (with a four-quantifier prefix) of a formula structure for which provability may be undecidable. Büchi (1962) showed, by exploiting Turing's methods further, that this proof could be simplified as well as the proofs of other important results, e.g. Trakhtenbrot's Theorem on the undecidability of the problem of validity in the class of all finite models.

2.3.2.2 Turing Machines and Word Problems

In his 1947 paper, Post proves the algorithmic unsolvability of Thue's Problem. This result (independently established by Markov) is noteworthy, for it seeks to give an answer to a classical problem that stems from algebra and not from internal issues of the theory of calculability or logic in general. It belongs indeed to the class of *word problems* for a given type of finitely presented algebraic structure (in the present case, semi-groups). Turing machines provide an essential link in Post's proof. They were later employed to establish similar unsolvability results for other structure types (semi-groups with cancellation in Turing 1950, Boone 1958, groups in Boone 1959). Words (or strings) are finite sequences of symbols from a given finite alphabet Σ. A system of n pairs of replacement rules, named 'productions', is considered:

$$(a_i) \quad \alpha A_i \beta \rightarrow \alpha B_i \beta \qquad (b_i) \quad \alpha B_i \beta \rightarrow \alpha A_i \beta \text{ (for } i = 1, \ldots, n)$$

where α and β are any words over Σ, and A_i and B_i are some fixed words over Σ.

These rules allow the substitution of B_i for a fixed occurrence of A_i in some word, or vice versa. Words U, V are said to be equivalent if V can be obtained from U by means of a finite sequence of applications of these productions. The problem, stated in 1914 by Thue (which we formulate here in Post's terminology), is the following: find a general method that, given an arbitrary system of the type just described, will determine, for any two arbitrary words U and V over Σ, whether they are equivalent or not, i.e. whether one of them is derivable from the other in this production system (Thue's general problem). In other words, Thue's Problem is the decision problem for derivability in Thue systems, which are a special sort among Post's canonical systems.

We call 'semi-Thue system' (STS) a system which only includes the first production (a_i) of each of the n pairs. There are STS with an unsolvable decision problem. Post's strategy is then to find such an STS having the property that the addition of the (b_i) productions allows no new assertion to be derived in the system.

Now this property is available when we consider a semi-Thue system obtained from a certain Turing machine (after a known unsolvable problem has been expressed in terms of this machine, as the reaching to a fixed state q_f). In fact, the computation of the machine reaching q_f when applied to a given word can be represented by a derivation of a fixed word V_0 in a system of semi-Thue productions; these productions are applied to words that represent successive complete configurations of the machine. Then a STS T_0 having an unsolvable decision problem is set up; its productions are the inverses of the latter ones, and its axiom is V_0. Thus the productions that have to be added for obtaining a Thue system T, are the productions representing the transitions of the machine. As a consequence of the deterministic nature of Turing machines, these productions can transform words derivable from V_0 in only one manner, namely, going back along some derivation already possible in T_0. Hence T has the same assertions as T_0, and its decision problem is undecidable. Thus in the fixed equivalence system corresponding to T, equivalence to the fixed word V_0 is undecidable. Solving Thue's general problem would therefore yield a solution to an unsolvable problem.

Post had used an unsolvable problem taken from his results on "normal systems". Kleene (1952, 382) gave a similar proof worked out from an undecidable predicate, which he had obtained in recursive function theory. These results illustrate what M. Davis has called the "unifying virtues" of the Turing machine formalism (Davis 1958, viii), by means of which connections can be established between different parts of computability theory.

2.3.3 Turing Machines and Automata Theory

The notion of a *finite automaton*, i.e. a device, whose capacity to store information consists in only a finite number of possible internal states, was somewhat suggested by Shannon (1938). Strictly speaking, it was presented in 1943 by McCulloch and Pitts, in the form of neural nets, but without the name and in a rather muddled way. J. von Neumann in his 1948 and 1949 suggested developing a "general and logical theory of automata" which could contribute to handling the problems of self-reproduction by using ideas particularly inspired by the universal Turing machine (von Neumann 1952–1953). In the book edited in 1956 by Shannon and J. McCarthy, *Automata Studies*, finite automata are clearly contrasted with infinite automata, namely Turing machines. Kleene, in his fundamental paper on finite automata (Kleene 1956, published in this book, but written in 1951), explicitly states that a Turing machine can be considered as a finite automaton supplied with an external, unbounded memory. Before that, finite automata were studied as finite systems built out from elementary components, each of which is in some way able to store elementary information: nets of formal neurons, or sequential circuits using electromagnetic or electronic components. From the work of Kleene, Moore, Mealy, and others, an *abstract* theory of automata began to develop, directly inspired by Turing's concept and which, like the latter, was based on the notion of an abstract state.

A finite automaton thus boils down to a strongly restricted Turing machine. Turing's representation of the computing device as a tape machine is adopted, but the reading head of the automaton can neither write nor erase nor go back. The word written on the tape is accepted if the sequence that results from scanning all successive symbols leads to an 'accepting' state. The role of the table of the Turing machine is here restricted to defining the state transition function, which is often given in the form of a graph (the transition diagram). This approach (where the finite automaton is usually regarded as a language recognizer) was especially developed by Rabin and Scott in their very rich paper 1959, the content of which, on a par with Kleene's theorem, constitutes the core of finite automata theory.

By means of Turing's model, Rabin and Scott could systematically investigate various ways of relaxing the initial restrictions, by allowing erasing or going backwards. They studied deterministic vs. nondeterministic, one-tape vs. two-tape finite automata. Thus their underlying concept of a machine was even more general than the usual concept of Turing machine. Finite automata with outputs can compute eventually periodic functions.

Such variations on the notion of Turing machine inspired the conception of other types of automata, related to the investigation of natural language grammars, programming languages and Post's combinatory systems. This gave rise to a hierarchy of automata, corresponding to Chomsky's hierarchy of grammar and language types, at the top of which Turing machines stand as an organizing concept for automata theory (Chomsky 1963).

2.3.4 The Study of the 'Classical' Turing Machine

Initiated by Turing's concept of a machine, automata theory pursued its investigations by systematically exploring the possible restrictions or generalizations of the Turing-Post-Kleene model. These machines have the same general structure as the original Turing machine, namely, that of a finite control machine operating sequentially on an infinite tape that can be used for storing information.

2.3.4.1 Quantitative Restrictions on Turing Machines

Submitting the Turing machine to what, on the surface, appear to be very strict quantitative or qualitative restrictions need not have devastating effects. For example, Shannon proved in 1956 that, for any Turing machine, there exists (a) an equivalent Turing machine using only two symbols, and (b) an equivalent Turing machine using only two states, provided the number of states (respectively, the number of symbols) is suitably increased. Even for the universal Turing machine, for example, no more than four symbols and seven states are needed. Moreover, universality "can be hidden in the details of operation and not clearly represented in the topology of the interstate connections" (Minsky 1967, 281).

All this combinatory work on the variants of a Turing machine confirmed its robustness and versatility. Machine simulation procedures that were studied on this occasion contributed to a sharper understanding of the inner workings of a Turing machine.

2.3.4.2 Multi-tape and Nondeterministic Turing Machines

The investigation of multi-tape and multi-head Turing machines did not aim to invalidate Turing's Thesis by exhibiting any more powerful machine. Its purpose was to produce a model of a machine that was closer to actual computers. Yet, in some cases, motivations were also theoretical ones. An undecidability result established by Rabin and Scott about two-tape, two-way finite automata gave rise to an interest in multi-tape machines: McCarthy conjectured that the source of this undecidability was a profound similarity between these automata and Turing machines, which Minsky, in 1961, proved to be the case.

The work by Rabin and Yamada in their studies on "real-time computation" prepared the way for computational complexity theory, which, since the mid 1960s, makes ample use of multi-tape Turing machines. For example, Hartmanis and Stearns showed in 1965 that a computation carried out in n units of time by a two-tape machine can require up to n^2 units of time to be carried out on a single-tape machine. But in complexity theory, usual single-tape Turing machines remain the convenient standard.

The Turing machine was confirmed in this role by a further generalization: the nondeterministic Turing machine (NDTM), i.e. a machine for which, given any configuration (state, scanned symbol), the transition function can take a finite, bounded number of values (triples consisting of a state, symbol to be printed and a displacement). A nondeterministic Turing machine, if used to recognize some language, accepts a word if and only if at least one of the possible computations available once the word is scanned leads to an accepting state. It is well known that the distinction between deterministic and nondeterministic algorithms is of prime importance in complexity theory today. In 1936, Turing had considered the possibility of nondeterministic machines, but only studied "automatic", that is to say, deterministic, machines. Apparently, the NDTM was not itself studied until 1960. Its equivalence to a deterministic Turing machine (disregarding the amount of time needed for the computation) was regarded as "well known", but was proved rigorously only in the late 1960s.

Using multi-tape machines to investigate the effects of restrictions on the elementary operations of Turing machines yielded a surprising result. In his 1961 proof of the unsolvability of Post's tag problem, Minsky showed that any Turing machine can be simulated by a two-tape, non-writing Turing machine. This machine is supplied with two, one-way infinite tapes whose squares are all empty, except for the first square of each tape, which bears a distinctive mark. In this machine, information is carried only by the position of the heads relative to the ends of their respective tapes, which is expressed by two numbers.

The proof and the very formulation of Minsky's theorem resort to a complex numerical coding of inputs and outputs. P. Fischer (1966), followed by Hopcroft and Ullman (1979), simplified the result by using extra tapes (distinct from the computation tapes) for inputs and outputs. Their proof rests on some simple lemmas concerning two types of Turing machines for which information storage on the tape is subject to strict constraints: the push-down automaton and the counter. The stack or push-down store is a one-way infinite tape standing up on end, where each symbol is stored one above the other on some initial stack start symbol, just as one would successively stack plates on some initial plate. Previously stored information is "pushed down" and only the last stored symbol can be read and modified or erased. The stack or push-down is a LIFO (Last In First Out) storing device. A counter is a rudimentary stack in which the machine can stack up 1's or remove them one by one. It can determine whether or not there are 1's in a stack, that is to say, it can store a natural number, increment or decrement it, and tell whether or not it is zero. The main tapes of Minsky's machine are counters: they can store and modify a number by the position of their heads.

Now (1) An arbitrary (single-tape) Turing machine can be simulated by a deterministic two-stack machine: one of the stacks is used to store the content of the part of the tape left of the head of the simulated machine, and the other for the part to the right of the head. (2) Two counters are enough to simulate one stack. Symbol manipulation on the stack is simulated by arithmetical operations on the content of one of the counters; the other is used to keep count of the operation loops to be carried out. Therefore a four-counter machine can simulate a Turing machine. (3) The content (a, b, c, d) of four counters can be represented by the number $n = 2^a 3^b 5^c 7^d$, which can be processed by a two-counter machine simulating the four-counter one. Thus Minsky's result is obtained. It has important applications for the study of Post's 'normal systems': namely, the unsolvability of the tag problem (which was Minsky's initial purpose), and the universality of monogenic normal systems.

It is noteworthy that the tapes of Minsky's machines, being counters, are nothing more than numerical registers that are managed by programs made from basic arithmetical instructions. So the removal of the writing function lead to a very abstract view of information processing on the tape of a Turing machine, and gradually suggested a new model of a machine, which also resulted from other research work done in the 1960s.

2.4 Reworking Turing's Model

From 1957 onwards, there have been various attempts to revise Turing's model of computation in a radical manner. The aim of these attempts was to bring computability theory and ordinary programming closer together, to get a more convenient framework to study the effects of various restrictions on the set of available operations, and to relate the theory of abstract machine computability more easily to other parts of the domain.

If 'program' is taken in a very large sense, a Turing machine table, or its description coded on the tape of the universal Turing machine, can be seen as a program. However, as C.Y. Lee remarks (1960), Turing's theory is essentially a theory of a special machine, based on the notion of state rather than on the notion of instruction. It does not provide a conjoint formalization of the notions of machine and program.

In a Turing machine, memory (the tape) is serial. Furthermore, quintuples which make up its table are just enumerated in an irrelevant ordering, and jump instructions are constantly used. On the other hand, the main memory of a computer is random-access (RAM), and a program is given, as much as possible, a sequential structure, for the programmer seeks to do away with or minimize breaks in instruction running (eliminating 'go to' instructions).

The transformation of Turing's model happened in two stages. The first one concerned *control structure*: states of special Turing machines are replaced by instructions organized in numbered sequences, i.e. *programs*. The second stage concerns *memory structure*: *registers* constituting RAM replace the tape of the Turing machine.

2.4.1 Program Machines and Hao Wang's Machine B

As early as 1936, Turing pointed out that, in his description of computation, it was possible to replace the "state of mind" of the (human) computer by a "note of instructions"; but he does not elaborate on the remark, which was probably only meant to show that his analysis of computation did not rest on doubtful psychological assumptions. Moreover, although the "sets of directions" in Post 1936 were close to programs, Post did not explicitly formulate any concept of machine on which these "sets of directions" could be implemented, nor (like Turing) did he give the constraints the "worker" who carries them out is subject to.

In 1954, H. Hermes proved that every Turing computable function can be computed by an idealized programmable computer. Hermes uses a "slightly modified" version of the Turing machine, one that is composed of five elementary machines. His computer is supplied with a finite number of registers that can store arbitrarily large numbers. Hermes's objective was to prove the "universality of programmable computers" rather than to replace the usual Turing machine by another model, but his approach bears certain similarities to the methods later used by Hao Wang and Minsky. Shortly afterwards, accounts of Turing machine begin to appear in which it is presented as the idealized version of a programmable, automatic, calculating machine: unlimited memory, maximal decomposition of operations, extremely simplified addressing mode (Trakhtenbrot 1957).

The most accomplished early attempt explicitly to introduce the concept of program into computability theory is due to Hao Wang (1957a). Wang's aim was twofold: on the one hand, to strengthen a classical result by using a new model of computation to prove that all partial recursive functions can be computed by *non-erasing* machines, and in so doing, to improve the tools available for undecidability proofs; on the other, by means of program machines, to elucidate the links between computability theory and computer practice.

Wang's paper is especially centered on the "Basic Machine B", a mix between a Turing machine and an idealized computer. From the Turing machine, it keeps the "serial storage", i.e. the (two-way infinite) tape, the reading-writing head and a control element. Furthermore, it also includes a "parallel storage" device, namely an internal, indefinitely extendable, random-access memory divided into cells, where any finite set of instructions can be stored: it is a programmable machine. When supplied with a program, the machine B is comparable to a special Turing machine. It uses only one symbol ($*$); as the machine cannot erase, there are only four basic operations: elementary movements to the right and to the left, print $*$, and jump to instruction n. A program is a finite set of instructions (ordered pairs) $(1, O_1), \ldots, (k, O_k)$, where each O_i is some elementary operation.

Strictly speaking, the machine B is not a universal Turing machine. The program it executes is not given to it in a coded form on its tape; the machine is supposed directly to interpret and execute the sequence of instructions stored in its parallel memory. This approach avoids all the complications related to the management, on the tape of the universal Turing machine, of the decoding and simulation of the imitated machine. On the other hand, as Wang recognizes, the machine B is highly fictional: "this assumption of an indefinite parallel storage whose units all are accessible at any moment is even more repulsive than permitting the tape to expand as needed" (Wang 1957a, 150).

Hao Wang obtains the aimed result. But he (rightly) conjectures that the distinction, in machine B, between two sorts of squares, which alternate on the tape, is not dispensable if erasing is not available. He also sketches a methodical study of systems of elementary instructions.

Wang's model of computation gives us the opportunity to reflect upon the constructive requirements that are appropriate for a theory of computing machines. Of course, like the Turing machine's tape, machine B's tape must never contain more than a finite number of printed squares (otherwise the machine would be equivalent to an oracle Turing machine, and could compute a function which is not Turing-computable). Wang also discusses the objection according to which an unlimited parallel storage is too fictional an assumption. One could transform the machine B into a universal Turing machine, which imitates any special Turing machine in a uniform way: a unique imitation program is stored in a finite, fixed parallel memory, and the tape is used for storing the program to be imitated in a coded form. If one can implant in the machine B a program that computes a universal element (i.e. some recursive function that enumerates all partial recursive functions), then this is enough to make it a 'functional' universal machine. A 'structural' universal machine, on the other hand, should imitate the detailed execution of any given program. In a second paper, Hao Wang wrote the program for such a machine after a painstaking "exercise in coding" (Wang 1957b). In this (non-erasing) machine, every step of the computation of the imitated machine is represented on the tape in a coded form.

Oberschelp, who was also inspired by ideas suggested by Hermes, completed these results in 1958. Asser (1959) and Kaphengst (1959) developed a model of an abstract computing machine, in which, as in machine B, "information memory" is separated from "program memory", with the aim of directly proving the equivalence between abstract computing machines and Markov's algorithms.

2.4.2 Diagrams and Registers

2.4.2.1 Three Sources of New Concepts of Computability: Actual Machines, Constructivism, Theory of (Normal) Algorithms

Wang's achievements concerned the control structure of the Turing machine. By reorganizing its memory structure, we were led to the program and register machine. The now classical version of it, proposed by Shepherdson and Sturgis (1963), was foreshadowed by work done especially by R. Péter (1958), Ershov (1958) and Kaphengst (1959). Péter introduced in 1958 what is usually called 'Kaluznin-Péter diagrams', because they merged when considering issues at the crossroads of Péter's constructivist concerns and a problem raised by Kaluznin. Kaluznin had set out to make precise the notion of flow-chart used in computer programming. He aimed to use this concept of "computability by diagrams" to establish a hierarchy of recursive functions according to the complexity of their diagrams. A Kaluznin-Péter (KP) diagram is a finite, directed graph with an associated set M. Mathematical vertices represent partial functions $M \rightarrow M$; logical vertices, where various tests can occur, represent partial functions $M \rightarrow \{T, F\}$. A diagram represents an operating process on M. The termination of the process after a finite number of steps determines a value, thus the diagram defines a partial function $M \rightarrow M$. Partial functions defined in this way can be associated to the mathematical vertices of further diagrams.

Péter considers a very restricted type of KP diagram, which she names "normal diagrams" (Normalschemata). The only initial logical functions they admit are identity tests, which can be applied to arbitrary n-uples of natural numbers; and the only initial mathematical functions they admit are given by a basis yielding the zero, successor and projection functions. As later emphasized by Shepherdson and Sturgis, although this is not wholly explicit in Péter's paper, these basic functions (that were directly suggested by recursive function theory) can be regarded as copying operations on finite sets of number registers (e.g. $+1$, *reset*, ...). Numerical data are stored on a finite (unbounded) number of registers. Function composition is ensured by the graphs.

Kaluznin (1959) tried to give a formulation of the concept of algorithm that avoided some of the practical drawbacks of Markov's normal algorithms: no representation of storage, intermediate words too long, complicated composition of algorithms. When interpreted in terms of word functions, normal KP diagrams are equivalent to normal algorithms.

Péter's interest in Kaluznin's problem is easily explained if we recall her well-known 1957 paper, "Rekursivität und Konstruktivität" (Péter 1959), in which she questioned the validity of Church's Thesis, or rather of its converse. In her opinion, truly finitely computable functions are only those for which a constructive definition is available. She readily admits that this is the case for all functions that are defined by special forms of recursion (even when these are not reducible to primitive recursion). On the contrary, she considers that the general recursive functions "can be said to be constructive only with a restriction that cannot be formulated without a vicious circle". Now Normalschemata seem to be rudimentary, truly constructive computing devices. We could conjecture that, by means of such a restricted class of diagrams, a proper sub class of partial recursive functions could be characterized which would nevertheless include every sort of special recursive function. But this attempt fails, as Péter proved in her paper on Graphschemata, because it turns out that every partial recursive function is computable by a Normalschema.

Ershov developed similar notions about the "schematic notation" of algorithms. Apparently, Kaphengst was the first, in 1959, explicitly to propose replacing the

Turing machine with a model of program and register machines. His "programmable computing machine", which he proved is universal, is an idealized computer having an infinite number of registers or "compartments" (Fächer): "computing register", instruction counter, cells where arbitrary finite sequences of 1's can be stored. These authors no longer distinguish two sorts of memory; they try to integrate the notion of stored program into their model of computation.

2.4.2.2 Shepherdson and Sturgis's Register Machines

Shepherdson and Sturgis's very important paper "Computability of recursive functions" was published in 1963, but the main ideas developed in it were known since 1960. Their strategy was to develop, in place of the Turing machine, a rich, realistic and flexible model of computation, without any worries about the economy of means, in which computing recursive functions could be easily programmed. Various restrictions on the set of instructions and on the number of registers can then be gradually imposed. In the course of this inquiry, the authors encounter and discuss most of the previous, aforementioned work, going back to Hermes' and Wang's papers.

These machines are controlled by programs that are seen as separately stored. The memory is infinite and is split up into registers; each register can store a number or word of arbitrary finite size. But there are different levels in the infinitary idealization.

The *Unlimited Register Machine* (URM) involves a countable set of number registers; but in any given program, only a finite number of them can occur and contain a non-zero number. The set of basic instructions (*add 1, subtract 1, clear, copy*, unconditional and conditional jump) is not minimal, but is selected for its convenience. The notion of a subroutine (with one or several exits) is carefully defined. It is quite easy to prove that every partial recursive function is computable by the URM. The set of basic instructions can be reduced to a simpler one, provided that many more registers are available. The URM can be directly extended to compute partial functions on words (on any given, finite alphabet) without resorting to arithmetization. It is thus rather similar to Post's normal systems.

The *Limited Register Machine* (LRM) only has a finite number of registers at any time, but instructions that add or remove registers are available. The equivalence of LRM to URM is easily proved.

'It is much more difficult to prove that the computational power of the URM (or the LRM) can be preserved in machines having a fixed, bounded number of registers or even *a single register* (SRM). The storage state of the LRM at any stage can be coded by a sequence of N numbers or words (namely, the contents of the N registers), or even by a single word, if an extra 'separator' symbol is added to the alphabet of the machine. But in order to simulate the basic operations of the LRM, it would be necessary to have other instructions that act on the components within the word. In order to avoid such complications, one can seek inspiration in Post's normal systems: a subroutine operates in such a way, that the parts to be modified occur only at the head or at the tail of the word. So every partial recursive function is computable by a SRM operating on the alphabet {0, 1} (where numbers are represented as sequences of 1's and 0 is a separator). Its single register can be seen as a two-headed, one-way tape: the reading head deletes the head of the word, and the printing takes place at the tail; it can also be seen as a 'stack' made up of 0's and 1's, writing being done only at the top, reading and deletion being done only at the bottom. This result suggests that access to both ends of the storage stack (which in this case is a FIFO (First In First Out) memory) is a very strong property. It is clearly related to Post's normal form theorem, which states that every system of productions is equivalent to some "normal system" (Post 1943).

The final reduction to a Turing machine is accomplished by writing subroutines equivalent to the basic instructions of the SRM and composed of the 'basic instructions' of Turing machines. Wang's results can also be obtained or strengthened.

The reduction to a SRM is possible even without extending the alphabet {1}; but in this case coding of numbers and basic operations need to be more complex. This is what Minsky's aforementioned 1961 results showed, when these are reinterpreted in terms of registers. In order to use the simpler basic operations mentioned earlier, we would require only one additional register. As stressed by Minsky (1967, 259), this latter result is 'better' than the former one, for an operation (even multiplication by 2) that requires an unbounded number of truly elementary actions cannot be considered a truly elementary one.

The major interest of Shepherdson and Sturgis' outstanding paper lies more in their systematic exploration of the connections between various machine models and their methodical reduction of the URM to the Turing machine than in making the idea of a program and register machine completely explicit, which was the subject of much discussion at the time. In 1961, a very similar model was presented in Canada: the Q-machine (Melzak) or "infinite abacus" (Lambek). It is a useful, intermediate link in the proof of the Turing- computability of every recursive function (cf. Boolos and Jeffrey 1974).

As Hartmanis (1981) later pointed out, talking about multi-tape machines, it would be illusory to believe that there is a unique satisfactory theoretical model of the computer. In his book (1974) intended for computer scientists, Manna uses, for various purposes, the classical Turing machine, the two-register (or two-stack) machine, and "Post's machine"; the latter can be understood as a tape machine subject to special constraints or as a machine with a single register arranged as a FIFO memory.

The style of such accounts reveals the influence of the views formulated by Dana Scott (1967). Scott requested that the theory clearly distinguish the notion of program ("a structured set of instructions") from that of machine. Turing's model does not fulfill this condition: in a special Turing machine, program and machine cannot be distinguished; in the universal Turing machine, the structure of the program to be executed is hidden under a complicated coding. A machine is specified by giving the indications needed to enter the data, execute the operations and tests, and extract the results. With respects to these demands, a register machine has the advantage of being much more elementary than a Turing machine.

Minsky's 1967 book, *Computation*, markedly popularized program and register machines. Matiyasevich (1984) shows a striking application of these machines, which he names 'Minsky machines'. He uses them (in a new version of his proof of the unsolvability of Hilbert's Tenth Problem) to give a new proof of an essential intermediate result established in 1961 by M. Davis, H. Putnam and J. Robinson, namely, the unsolvability of Hilbert's problem in the special case of exponential Diophantine equations. Here these machines are particularly convenient because they operate directly on natural numbers. Starting with a normal algorithm that has an undecidable halting problem, a Minsky machine can be specified which simulates it and which has an unsolvable halting problem too. A set of equations can then be given that simulates the operation of the machine. By several rather complicated transformations, a system of exponential Diophantine equations can be reduced to a system with a single equation. This equation, with parameters $A_0, \ldots, A_n$, admits a solution in natural numbers relative to the other variables if and only if, when the registers $R_0, \ldots, R_n$ are initially set to these numbers, the machine eventually halts. Thus the question is undecidable. By showing that $a = b^c$ can be expressed in terms of the existence of solutions for ordinary Diophantine equations, Matiyasevich completes the proof of unsolvability for the Tenth Problem.

2.5 The Limits of the Present Study

This paper did not attempt to give a complete survey of what could be called Turing's legacy in the modeling of computability. However, by duty, I must nonetheless explain why I neglected certain questions.

2.5.1 The Lack of Cellular Automata

Most of the models of computation considered in this paper are directly derived from the Turing machine, and were developed in the context of mainstream 'Automata Theory'. This work was achieved in the period from 1940 to 1965. On the other hand, I did not get into cellular automata and parallel computation models. Most of the investigations of cellular automata were done outside the aforementioned period. In the years 1948–1954, J. von Neumann had certainly formulated the concept of a universal constructor and had tried to set out a cellular ("crystalline", in his words) model of self-reproduction. But this work wasn't completed and published by A. Burks until 1966. Other investigations in this domain hardly took place before this date. Later, in the 1970s and 1980s, work by Conway, Toffoli and Wolfram gave a further impetus to the study of cellular automata (that is still going on today). As for the theory of parallel computing, it was mostly developed after 1970 (by Shepherdson for example).

2.5.2 *From Turing's "puzzles" to "K-graphs"*

Two noteworthy papers published in the 1950s were not mentioned: Turing 1954 and Kolmogorov and Uspensky 1958. The concept of a machine does not occur in the former and is seemingly not a central concern in the latter. Furthermore, both failed to arouse much interest at the time. R. Gandy nevertheless called attention to Kolmogorov and Uspensky's work, after which W. Sieg clearly showed the great theoretical importance of these papers, as both extensions and generalizations of the analysis of computability proposed by Turing in 1936. Since Sieg has already published, from 1990 onwards (Sieg 2009), many penetrating studies on the subject, I will refrain from going into the details and will only recall his main contributions to these questions.

In his informal 1954 paper, Turing does not deal with machines, but with 'puzzles'. He shows that applying mechanical procedures to tri-dimensional objects can be described by production systems in the style of Post. As explained by Sieg, Turing suggests that computation consists in operating on connected configurations containing one distinguished element, by substituting new neighborhoods for a neighborhood of this element. Moreover, the formal tools that allow us to treat the general question raised by Turing in 1954 can be found in Kolmogorov and Uspensky's paper. From these authors, Sieg and Byrnes drew the concept of K-graph. The 1958 paper develops ideas that Kolmogorov had formulated in the early 1950s (see Uspensky 1992) in order to formalize the notion of a mechanical procedure applied to constructive objects – the type of objects an algorithm is intended to operate on. They are connected graphs on which we operate by subgraph substitution.

2.6 Conclusion

It seemed useful to expound Shepherdson and Sturgis' approach to infinite machines, because it is an interesting strategy for the investigation of models of computability: starting with a very liberal and convenient model of computation, they gradually add restrictions on it in a finitary spirit. Concepts of computation and concepts of a machine need to be essentially finitary ones, but in order to get universality some 'amount of infiniteness' must be involved. The various models of machines I mentioned illustrate how many different levels there are in the infinitary idealization. In that respect, Hao Wang's remarks about machine B were already significant. The studies we considered improve our understanding of the role of infinity in machines, as well as the ways in which they use available means and can make up for the lack of some means by the use of others. By its rudimentary character, the Turing machine can seem at first to be at the very bottom of the (infinite machine) ladder, but some infinite-memory automata (e.g. push-down, counter) are strictly less powerful than it, because of some restrictions on the access they have to their infinite store or on the amount of it they are allowed to use to carry out a computation. Nevertheless, simply combining two of these devices is enough to recover all the power of a Turing machine and its universality. Thus, in various ways, the work of Hao Wang, R. Péter, Minsky, Shepherdson and Sturgis, and also that of Kolmogorov and Uspensky and later Sieg and Byrnes, has contributed in giving us a more general and abstract view of what applying mechanical procedures to "constructive objects" consists in.

References

Asser, G. (1959). Turing-Maschinen und Markowsche Algorithmen. *Zeitschrift für Mathematische Logik und Grundlagen der Mathematik, 5*, 346–365.

Bernays, P. (1958). Remarques sur le problème de la décision en logique élémentaire. In *Le raisonnement en mathématiques et en sciences expérimentales, Colloques Internationaux du CNRS, 70* (pp. 39–44). Paris: CNRS.

Boolos, G., & Jeffrey, R. (1974). *Computability and logic*. London: Cambridge University Press.

Boone, W. J. (1958). An analysis of Turing's 'the word problem in semi-groups with cancellation'. *Annals of Mathematics (Series 2), 67*, 195–202.

Boone, W. J. (1959). The word problem. *Annals of Mathematics (Series 2), 70*, 207–265.

Büchi, R. (1962). Turing machines and the Entscheidungsproblem. *Mathematische Annalen, 148*, 201–213.

Burks, A. W. (1966). Introduction. In J. von Neumann (1966).

Chomsky, N. (1963). Formal properties of grammars. In R. D. Luce, R. R. Bush, & E. Galanter (Eds.), *Handbook of mathematical psychology* (Vol. 2, chap. 12). New York: Wiley.

Church, A. (1937). Review of Turing (1936). *Journal of Symbolic Logic, 2*, 42–43.

Davis, M. (1958). *Computability and unsolvability*. New York: McGraw-Hill.

Davis, M. (Ed.). (1965). *The undecidable*. Hewlett: Raven Press.

Davis, M., Putnam, H., & Robinson, J. (1961). The decision problem for exponential Diophantine equations. *Annals of Matematics (Series 2), 74*(3), 425–436.

Ershov, A. P. (1958). Operatornye algorifmy [Algorithmes d'opérateurs]. *Doklady Akademii Nauk SSSR, 122*, 967–970 [*Automat. Express 1* (1959), 20–23].

Fischer, P. (1966). Turing machines with restricted memory access. *Information and Control, 9*(4), 364–379.

Gandy, R. (1980). Church's thesis and principles for mechanisms. In J. Barwise, H. J. Keisler, & K. Kunen (Eds.), *The Kleene symposium* (pp. 123–148). Amsterdam: North-Holland.

Gandy, R. (1988). The confluence of ideas in 1936. In Herken (1988) (pp. 55–111).

Gandy, R. (2001), Preface to "On computable numbers", pp. 9–17 in Turing (2001).

Gödel, K. (1934). On undecidable propositions of formal mathematical systems. In Gödel (1986) (pp. 346–371).

Gödel, K. (193?). [Undecidable Diophantine propositions]. In Gödel (1995) (pp. 164–175).

Gödel, K. (1986). In S. Feferman et al. (Eds.), *Gödel's collected works I*. New York: Oxford University Press.

Gödel, K. (1995). In S. Feferman et al. (Eds.), *Gödel's collected works III*. New York: Oxford University Press.

Hartmanis, J. (1981). Observations about the development of theoretical computer science. *Annals of the History of Computing, 3*(1), 42–51.

Hartmanis, J., & Stearns, R. E. (1965). On the computational complexity of algorithms. *Transactions AMS, 117*, 285–306.

Herken, R. (Ed.). (1988). *The universal Turing machine*. Oxford/New York: Oxford University Press (2nd ed., 1995, Wien/New York: Springer).

Hermes, H. (1954). Die Universalität programmgesteuerter Rechenmaschinen. *Mathematisch-Physikalische Semsterberichte (Göttingen), 4*, 42–53.

Hermes, H. (1961). *Aufzählbarkeit, Entscheidbarkeit, Berechenbarkeit*. Berlin/Göttingen/Heidelberg: Springer.
Hopcroft, J., & Ullman, J. D. (1979). *Introduction to automata theory, languages and computation*. Reading: Addison Wesley.
Kaluznin, L. A. (1959). Ob algoritmizacii matematicheskikh zadach [On algorithmization of mathematical problems]. *Problémy Kybérnétiki, 2*, 51–67.
Kaphengst, H. (1959). Eine abstrakte programmgesteuerte Rechenmaschine. *Zeitschrift für Mathematische Logik und Grundlagen der Mathematik, 5*, 366–379.
Kleene, S. C. (1952). *Introduction to metamathematics*. Princeton: Van Nostrand.
Kleene, S. C. (1956). Representation of events in nerve nets and finite automata. Project RAND research memorandum RM-704 (1951). In C. E. Shannon & J. McCarthy (Eds.), *Automata studies* (pp. 3–41). Princeton: Princeton University Press.
Kleene, S. C. (1967). *Mathematical logic*. New York: Wiley.
Kleene, S. C. (1981a). Origins of recursive function theory. *Annals of History of Computing, 3*(1), 52–67.
Kleene, S. C. (1981b). The theory of recursive functions approaching its centennial. *Bulletin of the American Mathematical Society (New Series), 5*, 43–61.
Kleene, S. C. (1989). The writing of *introduction to metamathematics*. In Th. Drucker (Ed.), *Perspectives on the history of mathematical logic*. Boston: Birkhäuser.
Kolmogorov, A., & Uspensky, V. (1958). On the definition of an algorithm. *Uspekhi Matematicheskikh Nauk XIII, 4*(82), 1–28 [*AMS Translations, 21, 2* (1963), 217–245].
Lambek, J. (1961). How to program an infinite abacus. *Canadian Mathematical Bulletin, 4*(3), 295–302.
Lee, C. Y. (1960). Automata and finite automata. *Bell System Technical Journal, 39*, 1267–1295.
Manna, Z. (1974). *Mathematical theory of computation*. New York: McGraw Hill.
Matiyasevich, Yu. V. (1984). (Am. Transl.), On investigations on some algorithmic problems in algebra and number theory. *Proceedings of Steklov Institute of Mathematics, 168*(3) (1986), 227–252.
McCulloch, W. S., & Pitts, W. (1943). A logical calculus of the ideas immanent in nervous activity. *Bulletin of Mathematical Biophysics, 5*, 115–133. (Reprinted in McCulloch (1965), *Embodiments of Mind*. Cambridge: MIT)
Melzak, Z. A. (1961). An informal arithmetical approach to computability and computation. *Canadian Mathematical Bulletin, 4*(3), 279–293.
Minsky, M. (1961). Recursive unsolvability of Post's problem of 'tag' and other topics in the theory of Turing machines. *Annals of Mathematics, 74*(3), 437–454.
Minsky, M. (1967). *Computation – Finite and infinite machines*. Englewood Cliffs: Prentice Hall.
von Neumann, J. (1948). The general and logical theory of automata. In *Cerebral mechanisms in behavior: The Hixon symposium* (pp. 1–41). New York: Wiley. (Reprinted in J. von Neumann (1963). *Collected works V*, pp. 288–328)
von Neumann, J. (1949). Theory and organization of complicated automata. In *Lectures at the University of Illinois* (pp. 31–87). In von Neumann (1966).
von Neumann, J. (1952–1953). The theory of automata: Construction, reproduction, homogeneity, part II. In *Theory of self-reproducing automata* (pp. 91–296). Urbana: University of Illinois Press.
von Neumann, J. (1963). *Collected works* (A. H. Taub, Ed., vol. V). Oxford/London/New York/Paris: Pergamon Press.
von Neumann, J. (1966). A. W. Burks (Ed. and completed), *Theory of self-reproducing automata*. Urbana: University of Illinois Press.
Oberschelp, W. (1958). Varianten von Turingmaschinen. *Archiv für Mathematische Logik und Grundlagenforschung, 4*, 53–62.
Péter, R. (1951). *Rekursive Funktionen*. Budapest: Akadémiai Kiado.
Péter, R. (1958). Graphschemata und rekursive Funktionen. *Dialectica, 12*, 373–393.
Péter, R. (1959), Rekursivität und Konstruktivität. In Heyting, A. (Ed.), *Constructivity in mathematics, proceedings of the colloquium*, Amsterdam, August 1957 (pp. 226–233). Amsterdam: North Holland.

Post, E. (1936). Finitary combinatory processes. Formulation 1. *Journal of Symbolic Logic, I*, 103–105.

Post, E. (1943). Formal reductions of the general combinatorial decision problem. *American Journal of Mathematics, 65*, 197–215.

Post, E. (1944). Recursively enumerable sets of positive integers and their decision problems. *Bulletin of the AMS, 50*, 284–316. (Reprinted in Davis 1965)

Post, E. (1947). Recursive unsolvability of a problem of Thue. *Journal of Symbolic Logic, 12*, 1–11. (Reprinted in Davis (1965), pp. 293–303)

Rabin, M. O., & Scott, D. (1959). Finite automata and their decision problems. *IBM Journal of Research and Development, 3*(2), 114–125. (Reprinted in E. F. Moore (Ed.), *Sequential machines* (pp. 63–91). Reading: Addison-Wesley)

Rogers, H., Jr. (1967). *Theory of recursive functions and effective computability*. New York: McGraw-Hill.

Scott, D. (1967). Some definitional suggestions for automata theory. *Journal of Computer and System Science, 1*(2), 187–212.

Shannon, C. E. (1938). A symbolic analysis of relay and switching circuits. *AIEE Transactions, 57*, 712–723.

Shannon, C. E. (1956). A universal Turing machine with two internal states. In C. E. Shannon & J. McCarthy (Eds.), *Automata studies* (pp. 157–165). Princeton: Princeton University Press.

Shannon, C. E., & McCarthy, J. (Eds.). (1956). *Automata studies*. Princeton: Princeton University Press.

Shepherdson, J. C., & Sturgis, H. E. (1963). Computability of recursive functions. *Journal of the ACM, 10* (1963), 217–255.

Sieg, W. (1994) Mechanical procedures and mathematical experience. In George, A. (Ed.), *Mathematics and mind* (pp. 71–117). Oxford: Oxford University Press.

Sieg, W. (2009). Computability theory. In A. Irvine, et al. (Eds.), *Handbook of the philosophy of science. Philosophy of mathematics*. Amsterdam/Boston: Elsevier.

Sieg, W., & Byrnes, J. (1996). K-graph machines: Generalizing Turing's machines and arguments. In P. Hajek (Ed.), *Gödel '96* (Lectures notes in logic 6, pp. 98–119). Berlin: Springer.

Sieg, W., & Byrnes, J. (1999). Gödel, Turing and K-graph machines. In A. Cantini, et al. (Eds.), *Logic and foundations of mathematics* (Synthese library, Vol. 280, pp. 57–66). Dordrecht: Kluwer.

Soare, R. I. (1996). Computability and recursion. *The Bulletin of Symbolic Logic, 2*(3), 284–321.

Thue, A. (1914). Probleme über Veränderungen von Zeichenreihen nach gegebenen Regeln (Skrifter utgit av Videnskappselskapet i Kristiania, I. Matematisk-naturvidenskabelig Klasse 1914 n 10). Kristiania: J. Dybwad.

Trakhtenbrot, B. A. (1957). *Algorifmi i machinnoie rechenie zadatch*. Moscou. 2d. publ., 1960. (Am. Transl., Boston: Heath, 1963. French Transl., *Algorithmes et machines à calculer*, Paris: Dunod, 1963)

Turing, A. M. (1936). On computable numbers, with an application to the Entscheidungsproblem. In *Proceedings of the London Mathematical Society (Series 2), 42*, 230–265. A correction, ibid., 43, 1937, 544–546. (Reprinted in Davis (1965), pp.115–154)

Turing, A. M. (1937). Computability and λ-definability. *Journal of Symbolic Logic, 2*, 153–163.

Turing, A. M. (1939). Systems of logic based on ordinals. *Proceedings of the London Mathematical Society (Series 2), 45*, 161–228. (Reprinted in Davis (1965), pp. 154–222)

Turing, A. M. (1950). The word problem in semi-groups with cancellation. *Annals of Mathematics (Series 2), 52*, 491–505.

Turing, A. M. (1954). Solvable and unsolvable problems. *Science News, 31*, 7–23. (Reprinted in Turing (1992), pp. 99–115)

Turing, A. M. (1992). J. L. Britton (Ed.), *Collected works: Pure mathematics*. Amsterdam: North Holland.

Turing, A. M. (2001). R. Gandy & C. Yates (Eds.), *Collected works: Mathematical logic*. Amsterdam: Elsevier.

Uspensky, V. A. (1992). Kolmogorov and mathematical logic. *Journal of Symbolic Logic, 57*, 383–412.

Wang, H. (1957a). A variant to Turing's theory of calculating machines. *Journal of the ACM, 4*(1), 63–92. (Reprinted in Wang, Hao (1970), ch. VI)

Wang, H. (1957b). Universal Turing machines: An exercise in coding. *Zeitschrift für Mathematische Logik und Grundlagen der Mathematik, 3*, 69–80. (Reprinted in Hao Wang (1970), ch. VII).

Wang, H. (1970). *Logic, computers and sets*. New York: Chelsea. (Reprint of Wang, H., *A survey of mathematical logic*, Peking 1962/New York 1964)

Wang, H. (1974). *From mathematics to philosophy*. London: Routledge/Kegan Paul.

Chapter 3
Kolmogorov Complexity in Perspective Part I: Information Theory and Randomness

Marie Ferbus-Zanda and Serge Grigorieff

Abstract We survey diverse approaches to the notion of information: from Shannon entropy to Kolmogorov complexity. Two of the main applications of Kolmogorov complexity are presented: randomness and classification. The survey is divided in two parts in the same volume. Part I is dedicated to information theory and the mathematical formalization of randomness based on Kolmogorov complexity. This last application goes back to the 1960s and 1970s with the work of Martin-Löf, Schnorr, Chaitin, Levin, and has gained new impetus in the last years.

Note. Following Robert Soare's recommendations in Soare (1996), which have now gained large agreement, we write *computable* and *computably enumerable* in place of the old fashioned *recursive* and *recursively enumerable (shorthand c.e.)*.

Notation. By $\log x$ (resp. $\log_s x$) we mean the logarithm of x in base 2 (resp. base s where $s \geq 2$). The "floor" and "ceil" of a real number x are denoted by $\lfloor x \rfloor$ and $\lceil x \rceil$: they are respectively the largest integer $\leq x$ and the smallest integer $\geq x$. Recall that, for $s \geq 2$, the length of the base s representation of an integer k is $\ell \geq 1$ if and only if $s^{\ell-1} \leq k < s^{\ell}$. Thus, the length of the base s representation of an integer k is $1 + \lfloor \log_s k \rfloor = 1 + \lfloor \frac{\log k}{\log s} \rfloor$.

The number of elements of a finite family $\mathcal{F}$ is denoted by $\sharp\mathcal{F}$.

The length of a word u is denoted by $|u|$.

3.1 Three Approaches to a Quantitative Definition of Information

A title borrowed from the seminal paper (Kolmogorov 1965).

M. Ferbus-Zanda • S. Grigorieff (✉)
LIAFA, CNRS & Université Paris Diderot – Paris 7, Case 7014, 75205 Paris Cedex 13, France
e-mail: Marie.Ferbus@liafa.univ-paris-diderot.fr; seg@liafa.univ-paris-diderot.fr

J. Dubucs, M. Bourdeau (eds.), *Constructivity and Computability in Historical and Philosophical Perspective*, Logic, Epistemology, and the Unity of Science 34,
DOI 10.1007/978-94-017-9217-2_3

3.1.1 *Which Information?*

3.1.1.1 About Anything...

About anything can be seen as conveying information. As usual in mathematical modelization, we retain only a few features of some real entity or process, and associate to them some finite or infinite mathematical objects. For instance,

- An integer or a rational number or a word in some alphabet, a finite sequence or a finite set of such objects, a finite graph, ...
- A real, a finite or infinite sequence of reals or a set of reals, a function over words or numbers, ...

This is very much as with probability spaces. For instance, to modelize the distributions of 6 balls into 3 cells, (cf. Feller 1968, §I.2, II.5) we forget everything about the nature of balls and cells and of the distribution process, retaining only two questions: "how many balls in each cell?" and "are the balls and cells distinguishable or not?". Accordingly, the modelization considers

- Either the $729 = 3^6$ maps from the set of balls into the set of cells in case the balls are distinguishable and so are the cells (this is what is done in Maxwell-Boltzman statistics),
- Or the $28 = \binom{6+(3-1)}{6}$ triples of non negative integers with sum[1] 6 in case the cells are distinguishable but not the balls (this is what is done in Bose-Einstein statistics)
- Or the 7 sets of at most 3 integers with sum 6 in case the balls are undistinguishable and so are the cells.

3.1.1.2 Especially Words

In information theory, special emphasis is made on information conveyed by words on finite alphabets. I.e., on *sequential information* as opposed to the obviously massively parallel and interactive distribution of information in real entities and processes. A drastic reduction which allows for mathematical developments (but also illustrates the Italian saying "traduttore, traditore!").

As is largely popularized by computer science, any finite alphabet with more than two letters can be reduced to one with exactly two letters. For instance, as exemplified by the ASCII code (American Standard Code for Information Interchange), any symbol used in written English – namely the lowercase and uppercase letters, the decimal digits, the diverse punctuation marks, the space, apostrophe, quote, left and

[1]For an easy proof, identify such a triple with a binary word with six letters 0 for the six balls and two letters 1 to mark the partition in the three cells.

right parentheses – together with some simple typographical commands – such as tabulation, line feed, carriage return or "end of file" – can be coded by binary words of length 7 (corresponding to the 128 ASCII codes). This leads to a simple way to code any English text by a binary word (which is 7 times longer).[2]

Though quite rough, the length of a word is the basic measure of its information content. Now, a fairness issue faces us: richer the alphabet, shorter the word. Considering groups of k successive letters as new letters of a super-alphabet, one trivially divides the length by k. For instance, a length n binary word becomes a length $\lceil \frac{n}{256} \rceil$ word with the usual packing of bits by groups of 8 (called bytes) which is done in computers. This is why all considerations about the length of words will always be developed relative to binary alphabets. A choice to be considered as a *normalization of length*.

Finally, we come to the basic idea to measure the information content of a mathematical object x:

information content of x	=	*length of a shortest binary word which "encodes" x*

What do we mean precisely by "encodes" is the crucial question. Following the trichotomy pointed in Kolmogorov (1965), we survey three approaches.

3.1.2 Combinatorial Approach: Entropy

3.1.2.1 Constant-Length Codes

Let us consider the family A^n of length n words in an alphabet A with s letters $a_1, \ldots, a_s$. Coding the a_i's by binary words w_i's all of length $\lceil \log s \rceil$, to any word u in A^n we can associate the binary word ξ obtained by substituting the w_i's to the occurrences of the a_i's in u. Clearly, ξ has length $n\lceil \log s \rceil$. Also, the map $u \mapsto \xi$ from the set A^* of words in alphabet A to the set $\{0, 1\}^*$ of binary words is very simple. Mathematically, considering on A^* and $\{0, 1\}^*$ the algebraic structure of monoid given by the concatenation product of words, this map $u \mapsto \xi$ is a morphism since the image of a concatenation uv is the concatenation of the images of u and v.

3.1.2.2 Variable-Length Prefix Codes

Instead of coding the s letters of A by binary words of length $\lceil \log s \rceil$, one can code the a_i's by binary words w_i's having different lengthes so as to associate short

[2] For other European languages with a lot of diacritic marks, one has to consider the 256 codes of Extended ASCII which have length 8. And for non European languages, one has to turn to the 65 536 codes of Unicode which have length 16.

codes to most frequent letters and long codes to rare ones. This is the basic idea of compression. Using such codes, the substitution of the w_i's to the occurrences of the a_i's in a word u gives a binary word ξ. And the map $u \mapsto \xi$ is again very simple. It is still a morphism from the monoid of words on alphabet A to the monoid of binary words and can also be computed by a finite automaton.

Now, we face a problem: can we recover u from ξ? i.e., is the map $u \mapsto \xi$ injective? In general the answer is no. However, a simple sufficient condition to ensure decoding is that the family $w_1, \ldots, w_s$ be a so-called *prefix-free code* (or *prefix code*). Which means that if $i \neq j$ then w_i is not a prefix of w_j.

This condition insures that there is a unique w_{i_1} which is a prefix of ξ. Then, considering the associated suffix ξ_1 of v (i.e., $v = w_{i_1}\xi_1$) there is a unique w_{i_2} which is a prefix of ξ_1, i.e., u is of the form $u = w_{i_1} w_{i_2} \xi_2$. And so on.

Suppose the numbers of occurrences in u of the letters $a_1, \ldots, a_s$ are $m_1, \ldots, m_s$, so that the length of u is $n = m_1 + \ldots + m_s$. Using a prefix-free code $w_1, \ldots, w_s$, the binary word ξ associated to u has length $m_1|w_1| + \ldots + m_s|w_s|$. A natural question is, given $m_1, \ldots, m_s$, *how to choose the prefix-free code* $w_1, \ldots, w_s$ *so as to minimize the length of* ξ *?*

Huffman (1952) found a very efficient algorithm (which has linear time complexity if the frequencies are already ordered). This algorithm (suitably modified to keep its top efficiency for words containing long runs of the same data) is nowadays used in nearly every application that involves the compression and transmission of data: fax machines, modems, networks,...

3.1.2.3 Entropy of a Distribution of Frequencies

The intuition of the notion of entropy in information theory is as follows. Given natural integers $m_1, \ldots, m_s$, consider the family $\mathcal{F}_{m_1,\ldots,m_s}$ of length $n = m_1 + \ldots + m_s$ words of the alphabet A in which there are exactly $m_1, \ldots, m_s$ occurrences of letters $a_1, \ldots, a_s$. How many binary digits are there in the binary representation of the number of words in $\mathcal{F}_{m_1,\ldots,m_s}$? It happens (cf. Proposition 3.1.2) that this number is essentially linear in n, the coefficient of n depending solely on the frequencies $\frac{m_1}{n}, \ldots, \frac{m_s}{n}$. It is this coefficient which is called the entropy H of the distribution of the frequencies $\frac{m_1}{n}, \ldots, \frac{m_s}{n}$.

Definition 3.1.1 (Shannon 1948). Let $f_1, \ldots, f_s$ be a distribution of frequencies, i.e., a sequence of reals in $[0, 1]$ such that $f_1 + \ldots + f_s = 1$. The entropy of $f_1, \ldots, f_s$ is the real

$$H = -(f_1 \log(f_1) + \ldots + f_s \log(f_s))$$

Proposition 3.1.2 (Shannon 1948). *Let* $m_1, \ldots, m_s$ *be natural integers and* $n = m_1 + \ldots + m_s$. *Then, letting* H *be the entropy of the distribution of frequencies* $\frac{m_1}{n}, \ldots, \frac{m_s}{n}$, *the number* $\sharp\mathcal{F}_{m_1,\ldots,m_s}$ *of words in* $\mathcal{F}_{m_1,\ldots,m_s}$ *satisfies*

$$\log(\sharp\mathcal{F}_{m_1,\dots,m_s}) = nH + O(\log n)$$

where the bound in $O(\log n)$ *depends solely on* s *and not on* $m_1, \dots, m_s$.

Proof. The set $\mathcal{F}_{m_1,\dots,m_s}$ contains $\frac{n!}{m_1! \times \dots \times m_s!}$ words. Using Stirling's approximation of the factorial function (cf. Feller 1968), namely $x! = \sqrt{2\pi}\ x^{x+\frac{1}{2}}\ e^{-x+\frac{\theta}{12}}$ where $0 < \theta < 1$, and equality $n = m_1 + \dots + m_S$, we get

$$\log\left(\frac{n!}{m_1! \times \dots \times m_s!}\right) = \left(\sum_i m_i\right)\log(n) - \left(\sum_i m_i \log m_i\right)$$
$$+\frac{1}{2}\log\left(\frac{n}{m_1 \times \dots \times m_s}\right) - (s-1)\log\sqrt{2\pi} + \alpha$$

where $|\alpha| \leq \frac{s}{12}\log e$. The difference of the first two terms is equal to $n[\sum_i \frac{m_i}{n}\log(\frac{m_i}{n})] = nH$ and the remaining sum is $O(\log n)$ since $n^{1-s} \leq \frac{n}{m_1 \times \dots \times m_s} \leq n$. □

H has a striking significance in terms of information content and compression. Any word u in $\mathcal{F}_{m_1,\dots,m_s}$ is uniquely characterized by its rank in this family (say relatively to the lexicographic ordering on words in alphabet A). In particular, the binary representation of this rank "encodes" u. Since this rank is $< \sharp\mathcal{F}_{m_1,\dots,m_s}$, its binary representation has length $\leq nH$ up to an $O(\log n)$ term. Thus, nH can be seen as an upper bound of the information content of u. Otherwise said, the n letters of u are encoded by nH binary digits. In terms of compression (nowadays so popular with the zip-like softwares), *u can be compressed to nH bits*, i.e., *the mean information content (which can be seen as the compression size in bits) of a letter of u is H*.

Let us look at two extreme cases.

- If all frequencies f_i are equal to $\frac{1}{s}$ then the entropy is $\log(s)$, so that the mean information content of a letter of u is $\log(s)$, i.e., there is no better (prefix-free) coding than that described in Sect. 3.1.2.1.
- In case some of the frequencies is 1 (hence all other ones being 0), the information content of u is reduced to its length n, which, written in binary, requires $\log(n)$ bits. As for the entropy, it is 0 (with the usual convention $0 \log 0 = 0$, justified by the fact that $\lim_{x\to 0} x\log x = 0$). The discrepancy between $nH = 0$ and the true information content $\log n$ comes from the $O(\log n)$ term in Proposition 3.1.2.

3.1.2.4 Shannon's Source Coding Theorem for Symbol Codes

The significance of the entropy explained above has been given a remarkable and precise form by Claude Elwood Shannon (1916–2001) in his celebrated paper (Shannon 1948). It's about the length of the binary word ξ associated to u via a prefix-free code. Shannon proved

- A lower bound of $|\xi|$ valid whatever be the prefix-free code $w_1, \ldots, w_s$,
- An upper bound, quite close to the lower bound, valid for particular prefix-free codes $w_1, \ldots, w_s$ (those making ξ shortest possible, for instance those given by Huffman's algorithm).

Theorem 3.1.3 (Shannon 1948). *Suppose the numbers of occurrences in u of the letters $a_1, \ldots, a_s$ are $m_1, \ldots, m_s$. Let $n = m_1 + \ldots + m_s$. Let H be the entropy of the considered distribution of frequencies $\frac{m_1}{n}, \ldots, \frac{m_s}{n}$.*

1. *For every prefix-free sequence of binary words $w_1, \ldots, w_s$ (which are to code the letters $a_1, \ldots, a_s$), the binary word ξ obtained by substituting w_i to each occurrence of a_i in u satisfies*

$$nH \leq |\xi|$$

2. *There exists a prefix-free sequence of binary words $w_1, \ldots, w_s$ such that*

$$nH \leq |\xi| < n(H + 1)$$

Proof. First, we recall two classical results.
Kraft's inequality. Let $\ell_1, \ldots, \ell_s$ be a finite sequence of integers. Inequality $2^{-\ell_1} + \ldots + 2^{-\ell_s} \leq 1$ holds if and only if there exists a prefix-free sequence of binary words $w_1, \ldots, w_s$ such that $\ell_1 = |w_1|, \ldots, \ell_s = |w_s|$.
Gibbs' inequality. Let $p_1, \ldots, p_s$ and $q_1, \ldots, q_s$ be two probability distributions, i.e., the p_i's (resp. q_i's) are in $[0, 1]$ and have sum 1. Then $-\sum p_i \log(p_i) \leq -\sum p_i \log(q_i)$ with equality if and only if $p_i = q_i$ for all i.
Proof of Point 1 of Theorem 3.1.3. Set $p_i = \frac{m_i}{n}$ and $q_i = \frac{2^{-|w_i|}}{S}$ where $S = \sum_i 2^{-|w_i|}$. Then

$$|\xi| = \sum_i m_i |w_i| = n \left[\sum_i \frac{m_i}{n}(-\log(q_i) - \log S) \right]$$

$$\geq n \left[-\left(\sum_i \frac{m_i}{n} \log\left(\frac{m_i}{n}\right) \right) - \log S \right] = n[H - \log S] \geq nH$$

The first inequality is an instance of Gibbs' inequality. For the last one, observe that $S \leq 1$.
Proof of Point 2 of Theorem 3.1.3. Set $\ell_i = \lceil -\log(\frac{m_i}{n}) \rceil$. Observe that $2^{-\ell_i} \leq \frac{m_i}{n}$. Thus, $2^{-\ell_1} + \ldots + 2^{-\ell_s} \leq 1$. Applying Kraft inequality, we see that there exists a prefix-free family of words $w_1, \ldots, w_s$ with lengthes $\ell_1, \ldots, \ell_s$. We consider the binary word ξ obtained via this prefix-free code, i.e., ξ is obtained by substituting w_i to each occurrence of a_i in u. Observe that $-\log(\frac{m_i}{n}) \leq \ell_i < -\log(\frac{m_i}{n}) + 1$. Summing, we get $nH \leq |\xi| < n(H + 1)$. □

In particular cases, the lower bound nH can be achieved.

Theorem 3.1.4. *In case the frequencies $\frac{m_i}{n}$'s are all negative powers of two (i.e., $\frac{1}{2}, \frac{1}{4}, \frac{1}{8}, \ldots$) then the optimal ξ (given by Huffman's algorithm) satisfies $\xi = nH$.*

3.1.2.5 Closer to the Entropy

In Sects. 3.1.2.3 and 3.1.2.4, we supposed the frequencies to be known and did not consider the information content of these frequencies. We now deal with that question.

Let us go back to the encoding mentioned at the start of Sect. 3.1.2.3. A word u in the family $\mathcal{F}_{m_1,\ldots,m_s}$ (of length n words with exactly $m_1, \ldots, m_s$ occurrences of $a_1, \ldots, a_s$) can be recovered from the following data:

- The values of $m_1, \ldots, m_s$,
- The rank of u in $\mathcal{F}_{m_1,\ldots,m_s}$ (relative to the lexicographic order on words).

We have seen (cf. Proposition 3.1.2) that the rank of u has a binary representation ρ of length $\leq nH + O(\log n)$. The integers $m_1, \ldots, m_s$ are encoded by their binary representations $\mu_1, \ldots, \mu_s$ which all have length $\leq 1 + \lfloor \log n \rfloor$. Now, to encode $m_1, \ldots, m_s$ and the rank of u, we cannot just concatenate $\mu_1, \ldots, \mu_s, \rho$: how would we know where μ_1 stops, where μ_2 starts,…, in the word obtained by concatenation? Several tricks are possible to overcome the problem, they are described in Sect. 3.1.2.6. Using Proposition 3.1.5, we set $\xi = \langle \mu_1, \ldots, \mu_s, \rho \rangle$ which has length $|\xi| = |\rho| + O(|\mu_1| + \ldots + |\mu_s|) = nH + O(\log n)$ (Proposition 3.1.5 gives a much better bound but this is of no use here). Then u can be recovered from ξ which is a binary word of length $nH + O(\log n)$. Thus, asymptotically, we get a better upper bound than $n(H + 1)$, the one given by Shannon for prefix-free codes (cf. Theorem 3.1.3).

Of course, ξ is no more obtained from u via a morphism (i.e., a map which preserves concatenation of words) between the monoid of words in alphabet A and that of binary words.

Notice that this also shows that prefix-free codes are not the only way to efficiently encode into a binary word ξ a word u from alphabet $a_1, \ldots, a_s$ for which the numbers $m_1, \ldots, m_s$ of occurrences of the a_i's are known.

3.1.2.6 Coding Finitely Many Words with One Word

How can we code two words u, v with only one word? The simplest way is to consider $u\$v$ where \$ is a fresh symbol outside the alphabet of u and v. But what if we want to stick to binary words? As said above, the concatenation of u and v does not do the job: how can one recover the prefix u in uv? A simple trick is to also concatenate the length of $|u|$ in unary and delimitate it by a zero. Indeed, denoting by 1^p the word $1\ldots1$ with p occurrences of 1, one can recover u and v from the word $1^{|u|}0uv$: the length of the first block of 1's tells where to stop in the suffix uv to get u. In other words, the map $(u, v) \to 1^{|u|}0uv$ is injective from $\{0,1\}^* \times \{0,1\}^* \to \{0,1\}^*$. In this way, the code of the pair (u, v) has length $2|u| + |v| + 1$. This can obviously be extended to more arguments using the map $(u_1, \ldots, u_s, v) \mapsto 1^{|u_1|}0^{|u_2|} \ldots \varepsilon^{|u_s|}\varepsilon' u_1 \ldots u_s v$ where $\varepsilon = 0$ if s is even and $\varepsilon = 1$ if s is odd and $\varepsilon' = 1 - \varepsilon$.

Proposition 3.1.5. *Let* $s \geq 1$. *There exists a map* $\langle\ \rangle : (\{0,1\}^*)^{s+1} \to \{0,1\}^*$ *which is injective and computable and such that, for all* $u_1, \ldots, u_s, v \in \{0,1\}^*$, $|\langle u_1, \ldots, u_s, v\rangle| = 2(|u_1| + \ldots + |u_s|) + |v| + 1$.

The following technical improvement will be needed in Part II §2.1.

Proposition 3.1.6. *There exists a map* $\langle\ \rangle : (\{0,1\}^*)^{s+1} \to \{0,1\}^*$ *which is injective and computable and such that, for all* $u_1, \ldots, u_s, v \in \{0,1\}^*$,

$$|\langle u_1, \ldots, u_s, v\rangle| = (|u_1| + \ldots + |u_s|) + (\log|u_1| + \ldots + \log|u_s|)$$
$$+2(\log\log|u_1| + \ldots + \log\log|u_s|) + |v| + O(1)$$

Proof. We consider the case $s = 1$, i.e., we want to code a pair (u, v). Instead of putting the prefix $1^{|u|}0$, let us put the binary representation $\beta(|u|)$ of the number $|u|$ prefixed by its length. This gives the more complex code: $1^{|\beta(|u|)|}0\beta(|u|)uv$ with length

$$|u| + |v| + 2(\lfloor\log|u|\rfloor + 1) + 1 \leq |u| + |v| + 2\log|u| + 3$$

The first block of ones gives the length of $\beta(|u|)$. Using this length, we can get $\beta(|u|)$ as the factor following this first block of ones. Now, $\beta(|u|)$ is the binary representation of $|u|$, so we get $|u|$ and can now separate u and v in the suffix uv. □

3.1.3 Probabilistic Approach: Ergodicity and Lossy Coding

The abstract probabilistic approach allows for considerable extensions of the results described in Sect. 3.1.2.

First, the restriction to fixed given frequencies can be relaxed. The probability of writing a_i may depend on what has already been written. For instance, Shannon's source coding theorem has been extended to the so called "ergodic asymptotically mean stationary source models".

Second, one can consider a lossy coding: some length n words in alphabet A are ill-treated or ignored. Let $0 < \delta < 1$ be the probability of this set of words. Shannon's theorem extends as follows:

- Whatever close to 0 is δ, one can compress u only down to nH bits,
- Whatever close to 1 is δ, one can achieve compression of u down to nH bits.

3.1.4 Algorithmic Approach: Kolmogorov Complexity

3.1.4.1 Berry's Paradox

So far, we considered two kinds of binary codings for a word u in alphabet $a_1, \ldots, a_s$. The simplest one uses variable-length prefix-free codes (Sect. 3.1.2.2). The other one codes the rank of u as a member of some set (Sect. 3.1.2.5). Clearly, there are plenty of other ways to encode any mathematical object. Why not consider all of them? And define the information content of a mathematical object x as *the shortest univoque description of x (written as a binary word)*. Though quite appealing, this notion is ill defined as stressed by Berry's paradox[3]:

> Let N be the *lexicographically least binary word which cannot be univoquely described by any binary word of length less than* 1000.

This description of N contains 106 symbols of written English (including spaces) and, using ASCII codes, can be written as a binary word of length $106 \times 7 = 742$. Assuming such a description to be well defined would lead to a univoque description of N in 742 bits, hence less than 1,000, a contradiction to the definition of N.

The solution to this inconsistency is clear: the quite vague notion of univoque description entering Berry's paradox is used both inside the sentence describing N and inside the argument to get the contradiction. A clash between two levels:

- The would be formal level carrying the description of N
- And the meta level which carries the inconsistency argument.

Any formalization of the notion of description should drastically reduce its scope and totally forbid any clash such as the above one.

3.1.4.2 The Turn to Computability

To get around the stumbling block of Berry's paradox and have a formal notion of description with wide scope, Andrei Nikolaievitch Kolmogorov (1903–1987) made an ingenious move: he turned to computability and replaced *description* by *computation program*. Exploiting the successful formalization of this a priori vague notion which was achieved in the 1930s.[4] This approach was first announced in Kolmogorov (1963) and then developed in Kolmogorov (1965). Similar approaches were also independently developed in Solomonov (1964a,b) and in Chaitin (1966, 1969).

---

[3] Berry's paradox is mentioned by Bertrand Russell (1908, p.222 or 150), who credited G.G. Berry, an Oxford librarian, for the suggestion.

[4] Through the works of Alonzo Church (via lambda calculus), Alan Mathison Turing (via Turing machines) and Kurt Gödel and Jacques Herbrand (via Herbrand-Gödel systems of equations) and Stephen Cole Kleene (via the recursion and minimization operators).

3.1.4.3 Digression on Computability Theory

The formalized notion of *computable function* (also called recursive function) goes along with that of *partial computable function* (also called partial recursive function) which should rather be called *partially computable partial function*, i.e., the *partial* character has to be distributed.[5] Thus, there are two theories:

- *The theory of computable functions,*
- *The theory of partial computable functions.*

The "right" theory, the one with a cornucopia of spectacular results, is that of partial computable functions. Let us pick up three fundamental results out of the cornucopia, which we state in terms of computers and programming languages. Let $\mathcal{I}$ and $\mathcal{O}$ be $\mathbb{N}$ or A^* where A is some finite or countably infinite alphabet (or, more generally, $\mathcal{I}$ and $\mathcal{O}$ can be elementary sets, cf. Definition 3.1.9).

Theorem 3.1.7. 1. [Enumeration theorem]. *The function which executes programs on their inputs:* (program, input) $\to$ output *is itself partial computable. This means that there exists a partial computable function*

$$U : \{0,1\}^* \times \mathcal{I} \to \mathcal{O}$$

such that the family of partial computable function $\mathcal{I} \to \mathcal{O}$ *is exactly* $\{U_e \mid e \in \{0,1\}^*\}$ *where* $U_e(x) = U(e,x)$. *Such a function* U *is called universal for partial computable functions* $\mathcal{I} \to \mathcal{O}$.

2. [Parameter theorem (or s^m_n thm)]. *One can exchange input and program* (this is von Neumann's key idea for computers).

   Formally, this means that, letting $\mathcal{I} = \mathcal{I}_1 \times \mathcal{I}_2$, *universal maps* $U_{\mathcal{I}_1 \times \mathcal{I}_2}$ *and* $U_{\mathcal{I}_2}$ *are such that there exists a computable total map* $s : \{0,1\}^* \times \mathcal{I}_1 \to \{0,1\}^*$ *such that, for all* $e \in \{0,1\}^*$, $x_1 \in \mathcal{I}_1$ *and* $x_2 \in \mathcal{I}_2$,

$$U_{\mathcal{I}_1 \times \mathcal{I}_2}(e,(x_1,x_2)) = U_{\mathcal{I}_2}(s(e,x_1),x_2)$$

3. [Kleene fixed point theorem]. *For any transformation of programs, there is a program which does the same* input $\to$ output *job as its transformed program.*[6] *Formally, this means that, for every partial computable map* $f : \{0,1\}^* \to \{0,1\}^*$, *there exists e such that*

$$\forall e \in \{0,1\}^* \quad \forall x \in \mathcal{I} \quad U(f(e),x) = U(e,x)$$

---

[5] In French, Lacombe (1960) used the expression *semi-fonction semi-récursive*.

[6] A seed for computer virology, cf. Bonfante et al. (2006).

3.1.4.4 Kolmogorov Complexity (or Program Size Complexity)

Turning to computability, the basic idea for Kolmogorov complexity[7] can be summed up by the following equation:

$$\boxed{\textit{description} \quad = \quad \textit{program}}$$

When we say "program", we mean a program taken from a family of programs, i.e., written in a programming language or describing a Turing machine or a system of Herbrand-Gödel equations or a Post system,... Since we are soon going to consider the length of programs, following what has been said in Sect. 3.1.1.2, we normalize programs: they will be binary words, i.e., elements of $\{0,1\}^*$.

So, we have to fix a function $\varphi : \{0,1\}^* \to \mathcal{O}$ and consider that the output of a program p is $\varphi(p)$.

Which φ are we to consider? Since we know that there are universal partial computable functions (i.e., functions able to emulate any other partial computable function modulo a computable transformation of programs, in other words, a compiler from one language to another), it is natural to consider universal partial computable functions. Which agrees with what has been said in Sect. 3.1.4.3.

Let us give the general definition of the Kolmogorov complexity associated to any function $\{0,1\}^* \to \mathcal{O}$.

Definition 3.1.8. If $\varphi : \{0,1\}^* \to \mathcal{O}$ is a partial function, set

$$K_\varphi : \mathcal{O} \to \mathbb{N} \quad , \quad K_\varphi(y) = \min\{|p| : \varphi(p) = y\}$$

with the convention that $\min \emptyset = +\infty$.

Intuition: p is a program (with no input), φ executes programs (i.e., φ is altogether a programming language plus a compiler plus a machinery to run programs) and $\varphi(p)$ is the output of the run of program p. Thus, for $y \in \mathcal{O}$, $K_\varphi(y)$ is the length of shortest programs p with which φ computes y (i.e., $\varphi(p) = y$).

As said above, we shall consider this definition for partial computable functions $\{0,1\}^* \to \mathcal{O}$. Of course, this forces to consider a set $\mathcal{O}$ endowed with a computability structure. Hence the choice of sets that we shall call *elementary* which do not exhaust all possible ones but will suffice for the results mentioned in this paper.

Definition 3.1.9. The family of elementary sets is obtained as follows:

- It contains $\mathbb{N}$ and the A^*'s where A is a finite or countable alphabet,
- It is closed under finite (non empty) product, product with any non empty finite set and the finite sequence operator.

[7] Delahaye's books (Delahaye 1999, 2006) present a very attractive survey on Kolmogorov complexity.

Note. Closure under the finite sequence operator is used to encode formulas in Theorem 3.2.2.

3.1.4.5 The Invariance Theorem

The problem with Definition 3.1.8 is that K_φ strongly depends on φ. Here comes a remarkable result, the invariance theorem, which insures that *there is a smallest* K_φ*, up to a constant.* It turns out that the proof of this theorem only needs the enumeration theorem and makes no use of the parameter theorem (usually omnipresent in computability theory).

Theorem 3.1.10 (Invariance theorem, (Kolmogorov 1965)). *Let $\mathbb{O}$ be an elementary set (cf. Definition 3.1.9). Among the K_φ's, where $\varphi : \{0,1\}^* \to \mathbb{O}$ varies in the family $PC^{\mathbb{O}}$ of partial computable functions, there is a smallest one, up to an additive constant (= within some bounded interval). I.e.*

$$\exists V \in PC^{\mathbb{O}} \quad \forall \varphi \in PC^{\mathbb{O}} \quad \exists c \quad \forall y \in \mathbb{O} \quad K_V(y) \le K_\varphi(y) + c$$

Such a V is called optimal. Moreover, any universal partial computable function $\{0,1\}^* \to \mathbb{O}$ is optimal.

Proof. Let $U : \{0,1\}^* \times \{0,1\}^* \to \mathbb{O}$ be partial computable and universal for partial computable functions $\{0,1\}^* \to \mathbb{O}$ (cf. point 1 of Theorem 3.1.7). Let $c : \{0,1\}^* \times \{0,1\}^* \to \{0,1\}^*$ be a total computable injective map such that $|c(e,x)| = 2|e|+|x|+1$ (cf. Proposition 3.1.5). Define $V : \{0,1\}^* \to \mathbb{O}$, with domain included in the range of c, as follows:

$$\forall e \in \{0,1\}^* \ \forall x \in \{0,1\}^* \quad V(c(e,x)) = U(e,x)$$

where equality means that both sides are simultaneously defined or not. Then, for every partial computable function $\varphi : \{0,1\}^* \to \mathbb{O}$, for every $y \in \mathbb{O}$, if $\varphi = U_e$ (i.e., $\varphi(x) = U(e,x)$ for all x, cf. point 1 of Theorem 3.1.7) then

$$\begin{aligned}
K_V(y) &= \text{least } |p| \text{ such that } V(p) = y \\
&\le \text{least } |c(e,x)| \text{ such that } V(c(e,x)) = y \\
&\qquad\qquad \text{(least is relative to } x \text{ since } e \text{ is fixed)} \\
&= \text{least } |c(e,x)| \text{ such that } U(e,x) = y \\
&= \text{least } |x| + 2|e| + 1 \text{ such that } \varphi(x) = y \\
&\qquad\qquad \text{since } |c(e,x)| = |x| + 2|e| + 1 \text{ and } \varphi(x) = U(e,x) \\
&= (\text{least } |x| \text{ such that } \varphi(x) = y) + 2|e| + 1 \\
&= K_\varphi(y) + 2|e| + 1
\end{aligned}$$

□

Using the invariance theorem, the Kolmogorov complexity $K^{\mathcal{O}} : \mathcal{O} \to \mathbb{N}$ is defined as K_V where V is any fixed optimal function. The arbitrariness of the choice of V does not modify drastically K_V, merely up to a constant.

Definition 3.1.11. Kolmogorov complexity $K^{\mathcal{O}} : \mathcal{O} \to \mathbb{N}$ is K_V, where V is some fixed optimal partial function $\{0,1\}^* \to \mathcal{O}$. When $\mathcal{O}$ is clear from context, we shall simply write K.

$K^{\mathcal{O}}$ is therefore minimum among the K_φ's, up to an additive constant.

$K^{\mathcal{O}}$ is defined up to an additive constant: if V and V' are both optimal then

$$\exists c \quad \forall x \in \mathcal{O} \quad |K_V(x) - K_{V'}(x)| \leq c$$

3.1.4.6 What Kolmogorov Said About the Constant

So Kolmogorov complexity is an integer defined up to a constant…! But the constant is uniformly bounded for $x \in \mathcal{O}$. Let us quote what Kolmogorov said about the constant in Kolmogorov (1965):

> Of course, one can avoid the indeterminacies associated with the [above] constants, by considering particular […functions V], but it is doubtful that this can be done without explicit arbitrariness.
>
> One must, however, suppose that the different "reasonable" [above optimal functions] will lead to "complexity estimates" that will converge on hundreds of bits instead of tens of thousands.
>
> Hence, such quantities as the "complexity" of the text of "War and Peace" can be assumed to be defined with what amounts to uniqueness.

In fact, this constant witnesses the multitude of models of computation: universal Turing machines, universal cellular automata, Herbrand-Gödel systems of equations, Post systems, Kleene definitions,… If we feel that one of them is canonical then we may consider the associated Kolmogorov complexity as the right one and forget about the constant. This has been developed for Schoenfinkel-Curry combinators S, K, I by Tromp, cf. Li and Vitányi (1997), §3.2.2–3.2.6.

However, even if we fix a particular K_V, the importance of the invariance theorem remains since it tells us that K is less than *any* K_φ (up to a constant). A result applied again and again to develop the theory.

3.1.4.7 Considering Inputs: Conditional Kolmogorov Complexity

Recall that, in computer science, inputs are also considered as *environments*. In the enumeration theorem, we considered *(program, input)* $\to$ *output* functions (cf. Theorem 3.1.7). Then, in the definition of Kolmogorov complexity, we gave up the inputs, dealing with *program* $\to$ *output* functions. Conditional Kolmogorov complexity deals with the inputs. Instead of measuring the information content of $y \in \mathcal{O}$, we measure it given as free some object z, which may help to compute y.

A trivial case is when $z = y$, then the information content of y given y is null. In fact, there is an obvious program which outputs exactly its input, whatever be the input.

Let us state the formal definition and the adequate invariance theorem.

Definition 3.1.12. If $\varphi : \{0,1\}^* \times \mathcal{I} \to \mathcal{O}$ is a partial function, set

$$K_\varphi(\ |\ ) : \mathcal{O} \times \mathcal{I} \to \mathbb{N} \quad , \quad K_\varphi(y \mid z) = \min\{|p| \mid \varphi(p,z) = y\}$$

Intuition: p is a program (with expects an input z), φ executes programs (i.e., φ is altogether a programming language plus a compiler plus a machinery to run programs) and $\varphi(p,z)$ is the output of the run of program p on input z. Thus, for $y \in \mathcal{O}$, $K_\varphi(y \mid z)$ is the length of shortest programs p for φ to compute y on input z, i.e., $\varphi(p,z) = y$.

Theorem 3.1.13 (Invariance theorem for conditional complexity). *Among the $K_\varphi(\ |\ )$'s, where φ varies in the family $PC^{\mathcal{O}}_{\mathcal{I}}$ of partial computable functions $\{0,1\}^* \times \mathcal{I} \to \mathcal{O}$, there is a smallest one, up to an additive constant (i.e., within some bounded interval):*

$$\exists V \in PC^{\mathcal{O}}_{\mathcal{I}} \quad \forall \varphi \in PC^{\mathcal{O}}_{\mathcal{I}} \quad \exists c \quad \forall y \in \mathcal{O} \quad \forall z \in \mathcal{I} \quad K_V(y \mid z) \leq K_\varphi(y \mid z) + c$$

Such a V is called optimal. Moreover, any universal partial computable map $\{0,1\}^* \times \mathcal{I} \to \mathcal{O}$ is optimal.

The proof is similar to that of Theorem 3.1.10.

Definition 3.1.14. $K^{\mathcal{I}\to\mathcal{O}} : \mathcal{O} \times \mathcal{I} \to \mathbb{N}$ is $K_V(\ |\ )$ where V is some fixed optimal partial function. It is defined up to an additive constant: if V et V' are both optimal then

$$\exists c \quad \forall y \in \mathcal{O} \quad \forall z \in \mathcal{I} \quad |K_V(y \mid z) - K_{V'}(y \mid z)| \leq c$$

Again, an integer defined up to a constant...! However, the constant is uniform in $y \in \mathcal{O}$ and $z \in \mathcal{I}$.

3.1.4.8 Simple Upper Bounds for Kolmogorov Complexity

Finally, let us mention rather trivial upper bounds:

- The information content of a word is at most its length.
- Conditional complexity is not harder than the non conditional one.

Proposition 3.1.15. *Let $f : \mathcal{O} \to \mathcal{O}'$ be computable. There exists c such that*

$$
\begin{array}{lr}
\forall x \in \{0,1\}^* & K^{\{0,1\}^*}(x) \leq |x| + c \\
\forall n \in \mathbb{N} & K^{\mathbb{N}}(n) \leq \log(n) + c \\
\forall x \in \mathcal{O} \quad \forall y \in \mathcal{I} & K^{\mathcal{I} \to \mathcal{O}}(x \mid y) \leq K^{\mathcal{O}}(x) + c \\
\forall x \in \mathcal{O} & K^{\mathcal{O}'}(f(x)) \leq K^{\mathcal{O}}(x) + c \\
\forall x \in \mathcal{O} \quad \forall y \in \mathcal{I} & K^{\mathcal{I} \to \mathcal{O}'}(f(x) \mid y) \leq K^{\mathcal{I} \to \mathcal{O}}(x \mid y) + c
\end{array}
$$

Proof. We only prove the first inequality. Let $Id : \{0,1\}^* \to \{0,1\}^*$ be the identity function. The invariance theorem insures that there exists c such that $K^{\{0,1\}^*} \leq K_{Id}^{\{0,1\}^*} + c$. Now, it is easy to see that $K_{Id}^{\{0,1\}^*} = |x|$, so that $K^{\{0,1\}^*}(x) \leq |x| + c$. Let $\theta : \{0,1\}^* \to \mathbb{N}$ be the bijection which associates to a word $u = a_{k-1} \dots a_0$ the predecessor of the integer with binary representation $1u$, i.e.,

$$\theta(u) = (2^k + a_{k-1}2^{k-1} + \dots + 2a_1 + a_0) - 1$$

Clearly, $K_{\theta}^{\mathbb{N}}(n) = \lfloor \log(n+1) \rfloor$. Use the invariance theorem to get c such that $K^{\mathbb{N}} \leq K_{\theta}^{\mathbb{N}} + c$. Then $K^{\mathbb{N}}(n) \leq \log(n) + c + 1$. □

The following technical property is a variation of an argument already used in Sect. 3.1.2.5: the rank of an element in a set defines this element, and if the set is computable, so is this process.

Proposition 3.1.16. *Let $A \subseteq \mathbb{N} \times \mathcal{O}$ be computable such that $A_n = A \cap (\{n\} \times \mathcal{O})$ is finite for all n. Then, letting $\sharp X$ be the number of elements of X,*

$$\exists c \quad \forall x \in A_n \quad K(x \mid n) \leq \log(\sharp(A_n)) + c$$

Proof. Observe that x is determined by its rank in A_n. This rank is an integer $< \sharp A_n$ hence its binary representation has length $\leq \lfloor \log(\sharp A_n) \rfloor + 1$. □

3.2 Kolmogorov Complexity

3.2.1 *Some Facts About K*

3.2.1.1 *K* Is Unbounded

Let $K = K_V : \mathcal{O} \to \mathbb{N}$ where $V : \{0,1\}^* \to \mathcal{O}$ is optimal (cf. Theorem 3.1.10). Since there are finitely many programs of size $\leq n$ (namely, the $2^{n+1} - 1$ binary words of size $\leq n$), there are finitely many elements of $\mathcal{O}$ with Kolmogorov complexity less than n. This shows that K is unbounded.

3.2.1.2 K Is Not Computable

Berry's paradox (cf. Sect. 3.1.4.1) has a counterpart in terms of Kolmogorov complexity: it gives a very simple proof that K, which is a total function $\mathcal{O} \to \mathbb{N}$, is not computable.

Proof that K is not computable. For simplicity of notations, we consider the case $\mathcal{O} = \mathbb{N}$. Define $L : \mathbb{N} \to \mathcal{O}$ as follows:

$$L(n) = \text{least } k \text{ such that } K(k) \geq 2n$$

So that $K(L(n)) \geq 2n$ for all n. If K were computable so would be L. Let $V:\mathcal{O} \to \mathbb{N}$ be optimal, i.e., $K = K_V$. The invariance theorem insures that there exists c such that $K \leq K_L + c$. Observe that $K_L(L(n)) \leq n$ by definition of K_L. Thus,

$$2n \leq K(L(n)) \leq K_L(L(n) + c \leq n + c$$

A contradiction for $n > c$. □

3.2.1.3 K and the Halting Problem

The non computability of K can be seen as a version of the undecidability of the halting problem. In fact, there is a simple way to compute K when the halting problem is used as an oracle. To get the value of $K(x)$, proceed as follows:

- Enumerate the programs in $\{0, 1\}^*$ in lexicographic order,
- For each program p, check if $V(p)$ halts (using the oracle),
- In case $V(p)$ halts then compute its value,
- Halt and output $|p|$ when some p is obtained such that $V(p) = x$.

The converse is also true: one can prove that *the halting problem is computable with K as an oracle.*

3.2.1.4 K Has No Computable Lower Bound

Though K is bounded from above by a total computable function, cf. Proposition 3.1.15, the argument for the undecidability of K can be used to prove that K is not bounded from below.

Theorem 3.2.1 (Kolmogorov). *There is no unbounded partial recursive function $\psi : \mathcal{O} \to \mathbb{N}$ such that $\psi(x) \leq K(x)$ for all x in domain(ψ).*

3.2.1.5 *K* Is Approximable from Above

Though K is not computable, it can be approximated from above. The idea is simple. Suppose $\mathcal{O} = \{0,1\}^*$. Let c be as in point 1 of Proposition 3.1.15. Consider all programs of length less than $|x| + c$ and let them be executed during t steps. If none of them converges and outputs x then take $|x| + c$ as a t-bound. If some of them converges and outputs x then the bound is the length of the shortest such program. The limit of this process is $K(x)$, it is obtained at some finite step which we are not able to bound.

Formally, this means that there is some $F : \mathcal{O} \times \mathbb{N} \to \mathbb{N}$ which is computable and decreasing in its second argument such that

$$K(x) = \lim_{t \to +\infty} F(x,t) = \min\{F(x,t) \mid t \in \mathbb{N}\}$$

3.2.2 *K and Gödel's Incompleteness Theorem*

A striking version of Gödel's incompleteness theorem has been given in Chaitin (1971, 1974), in terms of Kolmogorov complexity. Since Gödel's celebrated proof of the incompleteness theorem, we know that, in the language of arithmetic, one can formalize computability and logic. In particular, one can formalize Kolmogorov complexity and statements about it. Chaitin proves a version of the incompleteness theorem which insures that among true unprovable formulas there are all true statements $K(u) > n$ for n large enough.

Theorem 3.2.2 (Chaitin 1974). *Let $\mathcal{T}$ be a computably enumerable set of axioms in the language of arithmetic. Suppose that all axioms in $\mathcal{T}$ are true in the standard model of arithmetics with base $\mathbb{N}$. Then there exists N such that if $\mathcal{T}$ proves $K(u) > n$ (with $u \in \{0,1\}^*$ and $n \in \mathbb{N}$) then $n \leq N$.*

How the constant N depends on $\mathcal{T}$ has been giving a remarkable analysis by Chaitin. To that purpose, he extends Kolmogorov complexity to computably enumerable sets.

Definition 3.2.3 (Chaitin 1974). Let $\mathcal{O}$ be an elementary set (cf. Definition 3.1.9) and $\mathcal{CE}$ be the family of computably enumerable (c.e.) subsets of $\mathcal{O}$. If $\varphi : \{0,1\}^* \times \mathbb{N} \to \mathcal{O}$ is partial computable, let $K_\varphi : \mathcal{CE} \to \mathbb{N}$ be the Kolmogorov complexity such that, for all c.e. subset $\mathcal{T}$ of $\mathcal{O}$,

$$K_\varphi(\mathcal{T}) = \min\{|p| \mid \mathcal{T} = \{\varphi(p,t) \mid t \in \mathbb{N}\}\}$$

(observe that $\{\varphi(p,t) \mid t \in \mathbb{N}\}$ is always c.e. and any c.e. subset of $\mathcal{O}$ can be obtained in this way for some φ).

The invariance theorem still holds for this notion of Kolmogorov complexity, leading to the following notion.

Definition 3.2.4 (Chaitin 1974). $K^{\mathcal{CE}} : \mathcal{CE} \to \mathbb{N}$ is K_φ where φ is some fixed optimal partial function. It is defined up to an additive constant.

We can now state how the constant N in Theorem 3.2.2 depends on the theory $\mathcal{T}$.

Theorem 3.2.5 (Chaitin 1974). *There exists a constant c such that, for all c.e. sets $\mathcal{T}$ satisfying the hypothesis of Theorem 3.2.2, the associated constant N is such that*

$$N \leq K^{\mathcal{CE}}(\mathcal{T}) + c$$

Chaitin also reformulates Theorem 3.2.2 as follows:

> If $\mathcal{T}$ consist of true formulas then it cannot prove that a string has Kolmogorov complexity greater than the Kolmogorov complexity of $\mathcal{T}$ itself (up to a constant independent of $\mathcal{T}$).

Remark 3.2.6. The previous statement, and Chaitin's assertion that the Kolmogorov complexity of $\mathcal{T}$ somehow measures the power of $\mathcal{T}$ as a theory, has been much criticized in van Lambalgen (1989), Fallis (1996) and Raatikainen (1998). The main argument in Raatikainen (1998) against Chaitin's interpretation is that the constant in Theorem 3.2.2 strongly depends on the choice of the optimal function V such that $K = K_V$. Indeed, for any fixed theory $\mathcal{T}$, one can choose such a V so that the constant is zero! And also choose V so that the constant is arbitrarily large. Though these arguments are perfectly sound, we disagree with the criticisms issued from them. Let us detail three main rebuttals.

- First, such arguments are based on the use of optimal functions associated to very unnatural universal functions V (cf. point 1 of Theorem 3.1.7 and the last assertion of Theorem 3.1.10). It has since been recognized that universality is not always sufficient to get smooth results. Universality by prefix adjunction is sometimes required, (cf., for instance, §2.1 and §6 in Becher et al. 2006). This means that, for an enumeration $(\varphi_e)_{e\in\{0,1\}^*}$ of partial computable functions, the optimal function V is to satisfy equality $V(ep) = \varphi_e(p)$, for all e, p, where ep is the concatenation of the strings e and p.
- Second, and more important than the above technical counterargument, observe that modelization rarely rules out all pathological cases. It is intended to be used in "reasonable" cases. Of course, this may be misleading, but perfect modelization is illusory. In our opinion, this is best illustrated by Kolmogorov's citation quoted in Sect. 3.1.4.6 to which Raatikainen's argument could be applied mutatis mutandis: there are optimal functions for which the complexity of the text of "War and Peace" is null and other ones for which it is arbitrarily large. Nevertheless, this does not prevent Kolmogorov to assert (in the founding paper of the theory): *[For] "reasonable" [above optimal functions], such quantities as the "complexity" of the text of "War and Peace" can be assumed to be defined with what amounts to uniqueness.*
- Third, a final technical answer to such criticisms has been recently provided in Calude and Jürgensen (2005). They improve the incompleteness result of

Theorem 3.2.2, proving that, for a class of formulas in the vein of those in that theorem, the probability that such a formula of length n is provable tends to zero when n tends to infinity whereas the probability that it be true has a strictly positive lower bound.

3.3 Kolmogorov Complexity: Some Variations

Note. The denotations of (plain) Kolmogorov complexity (that of Sect. 3.1.4.5) and its prefix version (cf. Sect. 3.3.3) may cause some confusion. They long used to be respectively denoted by K and H in the literature. But in their book (Li and Vitányi 1997), Li and Vitanyi respectively denoted them by C and K. Due to the large success of this book, these last denotations are since used in many papers. So that two incompatible denotations now appear in the literature. In this paper, we stick to the traditional denotations K and H.

3.3.1 Levin Monotone Complexity

Kolmogorov complexity is non monotone, be it on $\mathbb{N}$ with the natural ordering or on $\{0,1\}^*$ with the lexicographic ordering. In fact, for every n and c, there are strings of length n with complexity $\geq n(1-2^{-c})$ (cf. Proposition 3.4.2). However, since $n \mapsto 1^n$ is computable, $K(1^n) \leq K(n) + O(1) \leq \log n + O(1)$ (cf. point 3 of Proposition 3.1.15) is much less than $n(1-2^{-c})$ for n large enough.

Leonid Levin, introduced a monotone version of Kolmogorov complexity (Levin 1973). The idea is to consider possibly infinite computations of Turing machines which never erase anything on the output tape. Such machines have finite or infinite outputs and compute total maps $\{0,1\}^* \to \{0,1\}^{\leq\omega}$ where $\{0,1\}^{\leq\omega} = \{0,1\}^* \cup \{0,1\}^{\mathbb{N}}$ is the family of finite or infinite binary strings. These maps can also be viewed as limit maps $p \to \sup_{t\to\infty} \varphi(p,t)$ where $\varphi : \{0,1\}^* \times \mathbb{N} \to \{0,1\}^*$ is total monotone non decreasing in its second argument.

To each such map φ, Levin associates a monotone non decreasing map $K^{mon}_\varphi : \{0,1\}^* \to \mathbb{N}$ such that

$$K^{mon}_\varphi(x) = \min\{|p| \mid \exists t\ x \leq_{pref} \varphi(p,t)\}$$

Theorem 3.3.1 (Levin 1973).

1. *If φ is total computable and monotone non decreasing in its second argument then $K^{mon}_\varphi : \{0,1\}^* \to \mathbb{N}$ is monotone non decreasing:*

$$x \leq_{pref} y \Rightarrow K^{mon}_\varphi(x) \leq K^{mon}_\varphi(y)$$

2. *Among the K^{mon}_φ's, φ total computable monotone non decreasing in its second argument, there exists a smallest one, up to a constant.*

Considering total φ's in the above theorem is a priori surprising since there is no computable enumeration of total computable functions and the proof of the Invariance Theorem 3.1.10 is based on the enumeration theorem (cf. Theorem 3.1.7). Here is the trick to overcome that problem.

- Consider all partial computable $\varphi : \{0,1\}^* \times \mathbb{N} \to \{0,1\}^*$ which are total monotone non decreasing in their second argument.
- Associate to each such φ a total $\tilde{\varphi}$ defined as follows: $\tilde{\varphi}(p,t)$ is the largest $\varphi(p,t')$ such that $t' \leq t$ and $\varphi(p,t')$ is defined within $t+1$ computation steps if there is such a t'. If there is none then $\tilde{\varphi}(p,t)$ is the empty word.
- Observe that $K^{mon}_{\varphi}(x) = K^{mon}_{\tilde{\varphi}}(x)$.

In Sect. 3.5.2.3, we shall see some remarkable property of Levin monotone complexity K^{mon} concerning Martin-Löf random reals.

3.3.2 Schnorr Process Complexity

Another variant of Kolmogorov complexity has been introduced by Klaus Peter Schnorr in Schnorr (1973). It is based on the subclass of partial computable functions $\varphi : \{0,1\}^* \to \{0,1\}^*$ which are monotone non decreasing relative to the prefix ordering:

(*) $(p \leq_{pref} q \wedge \varphi(p), \varphi(q) \text{ are both defined}) \Rightarrow \varphi(p) \leq_{pref} \varphi(q)$

Why such a requirement on φ? The reason can be explained as follows.

- Consider a sequential composition (i.e., a pipeline) of two processes, formalized as two functions f, g. The first one takes an input p and outputs $f(p)$, the second one takes $f(p)$ as input and outputs $g(f(p))$.
- Each process is supposed to be monotone: the first letter of $f(p)$ appears first, then the second one, etc. Idem with the digits of $g(q)$ for any input q.
- More efficiency is obtained if one can develop the computation of g on input $f(p)$ as soon as the letters of $f(p)$ appear. More precisely, suppose the prefix q of $f(p)$ has already appeared but there is some delay to get the subsequent letters. Then we can compute $g(q)$. But this is useful only in case the computation of $g(q)$ is itself a prefix of that of $g(f(p))$. This last condition is exactly the requirement $(*)$.

An enumeration theorem holds for the φ's satisfying $(*)$, allowing to prove an invariance theorem and to define a so-called process complexity $K^{proc} : \{0,1\}^* \to \mathbb{N}$. Schnorr process complexity has many properties in common with Levin's monotone complexity, cf. Sect. 3.5.2.3.

3.3.3 *Prefix (or Self-Delimited) Complexity*

Levin (1974), Gács (1974) and Chaitin (1975) introduced the most successful variant of Kolmogorov complexity: the prefix complexity. The idea is to restrict the family of partial computable functions $\{0,1\}^* \to \mathcal{O}$ (recall $\mathcal{O}$ denotes an elementary set in the sense of Definition 3.1.9) to those which have prefix-free domains, i.e. any two words in the domain are incomparable with respect to the prefix ordering.

An enumeration theorem holds for the φ's satisfying $(*)$, allowing to prove an invariance theorem and to define the so-called prefix complexity $H : \{0,1\}^* \to \mathbb{N}$ (not to be confused with the entropy of a family of frequencies, cf. Sect. 3.1.2.3).

Theorem 3.3.2. *Among the K_φ's, where $\varphi : \{0,1\}^* \to \mathcal{O}$ varies over partial computable functions with prefix-free domain, there exists a smallest one, up to a constant. This smallest one (defined up to a constant), denoted by $H^{\mathcal{O}}$, is called the prefix complexity.*

This prefix-free condition on the domain may seem rather technical. A conceptual meaning of this condition has been given by Chaitin in terms of self-delimitation.

Proposition 3.3.3 (Chaitin 1974). *A partial computable function $\varphi : \{0,1\}^* \to \mathcal{O}$ has prefix-free domain if and only if it can be computed by a Turing machine $\mathcal{M}$ with the following property:*

> *If x is in domain(φ) (i.e., $\mathcal{M}$ on input p halts in an accepting state at some step) then the head of the input tape of $\mathcal{M}$ reads entirely the input p but never moves to the cell right to p.*

This means that p, interpreted as a program, has no need of external action (as that of an end-of-file symbol) to know where it ends: as Chaitin says, the program is self-delimited. A comparison can be made with biological phenomena. For instance, the hand grows during childhood and then stops growing. No external action prevents the hand to go on growing. There is something inside the genetic program which creates a halting signal so that the hand stops growing.

The main reason for the success of the prefix complexity is that, with prefix-free domains, one can use the Kraft-Chaitin inequality (cf. the proof of Theorem 3.1.3) and get remarkable properties.

Theorem 3.3.4 (Kraft-Chaitin inequality). *A sequence (resp. computable sequence) $(n_i)_{i\in\mathbb{N}}$ of non negative integers is the sequence of lengths of a prefix-free (resp. computable) family of words $(u_i)_{i\in\mathbb{N}}$ if and only if $\sum_{i\in\mathbb{N}} 2^{-n_i} \leq 1$.*

Let us state the most spectacular property of the prefix complexity.

Theorem 3.3.5 (The Coding Theorem (Levin 1974)). *Consider the family $\ell_1^{c.e.}$ of sequences of non negative real numbers $(r_x)_{x\in\mathcal{O}}$ such that*

- $\sum_{x\in\mathcal{O}} r_x < +\infty$ *(i.e., the series is summable),*
- $\{(x,q) \in \mathcal{O} \times \mathbb{Q} \mid q < r_x\}$ *is computably enumerable (i.e., the r_x's have c.e. left cuts in the set of rational numbers $\mathbb{Q}$ and this is uniform in x).*

The sequence $(2^{-H^{\mathcal{O}}(x)})_{x\in\mathcal{O}}$ *is in* $\ell_1^{c.e.}$ *and, up to a multiplicative factor, it is the largest sequence in* $\ell_1^{c.e.}$*. This means that*

$$\forall (r_x)_{x\in\mathcal{O}} \in \ell_1^{c.e.} \quad \exists c \quad \forall x \in \mathcal{O} \quad r_x \leq c\, 2^{-H^{\mathcal{O}}(x)}$$

In particular, consider a countably infinite alphabet A. Let $V : \{0,1\}^* \to A$ be a partial computable function with prefix-free domain such that $H^A = K_V$. Consider the prefix code $(p_a)_{a\in A}$ such that, for each letter $a \in A$, p_a is a shortest binary string such that $V(p_a) = a$. Then, for every probability distribution $P : A \to [0,1]$ over the letters of the alphabet A, which is computably approximable from below (i.e., $\{(a,q) \in A \times \mathbb{Q} \mid q < P(a)\}$ is computably enumerable), we have

$$\forall a \in A \quad P(a) \leq c\, 2^{-H^A(a)}$$

for some c which depends on P but not on $a \in A$. This inequality is the reason why the sequence $(2^{-H^A(a)})_{a\in A}$ is also called *the universal a priori probability* (though, strictly speaking, it is not a probability since the $2^{-H^A(a)}$'s do not sum up to 1).

3.3.4 Oracular and Sub-oracular Kolmogorov Complexity

3.3.4.1 Oracular Kolmogorov Complexity

As is always the case in computability theory, everything relativizes to any oracle Z. Relativization modifies the equation given at the start of Sect. 3.1.4.4, which is now

description $=$ *program of a partial* Z*-computable function*

and for each possible oracle Z there exists a Kolmogorov complexity relative to oracle Z. We shall see in Sect. 3.5.5.2 an interesting property involving oracular Kolmogorov complexity.

Oracles in computability theory can also be considered as second-order arguments of computable or partial computable *functionals*. The same holds with oracular Kolmogorov complexity: the oracle Z can be seen as a second-order condition for a *second-order conditional Kolmogorov complexity*

$$K(y \mid Z) \quad \text{where} \quad K(\ \mid\ ) : \mathcal{O} \times P(\mathcal{I}) \to \mathbb{N}$$

Which has the advantage that the unavoidable constant in the "up to a constant" properties does not depend on the particular oracle. It depends solely on the considered functional.

Finally, one can mix first-order and second-order conditions, leading to a conditional Kolmogorov complexity with both first-order and second-order conditions

$$K(y \mid z, Z) \quad \text{where} \quad K(\ \mid\ ,\ ) : \mathcal{O} \times \mathcal{I} \times P(\mathcal{I}) \to \mathbb{N}$$

3.3.4.2 Sub-oracular Kolmogorov Complexity

Going back to the idea of possibly infinite computations as in Sect. 3.3.1, let us define

$$K^{\infty} : \{0,1\}^* \to \mathbb{N} \quad , \quad K^{\infty}(x) = \min\{|p| \mid U(p) = x\}$$

where U is the map $\{0,1\}^* \to \{0,1\}^{\leq\omega}$ computed by a universal Turing machine with possibly infinite computations. This complexity lies between K and $K(\ \mid \emptyset')$ (where $\emptyset'$ is a computably enumerable set which encodes the halting problem):

$$\forall x \quad K(x \mid \emptyset') \leq K^{\infty}(x) + O(1) \leq K(x) + O(1)$$

This complexity is studied in Becher et al. (2005) and in our paper (Ferbus-Zanda and Grigorieff 2006).

3.4 Formalization of Randomness: Finite Objects

3.4.1 Sciences of Randomness: Probability and Cryptography

3.4.1.1 Probability Theory

Random objects *(words, integers, reals,...)* constitute the basic intuition for probabilities ... *but they are not considered per se.* No formal definition of random object is given: there seems to be no need for such a formal concept. The existing formal notion of *random variable* has nothing to do with randomness: a random variable is merely a *measurable function* which can be as non random as one likes. It sounds strange that the mathematical theory which deals with randomness removes the natural basic questions:

- *What is a random string?*
- *What is a random infinite sequence?*

When questioned, people in probability theory agree that they skip these questions but do not feel sorry about it. As it is, the theory deals with laws of randomness and is so successful that it can do without entering this problem.

This may seem to be analogous to what is the case in geometry. What are points, lines, planes? No definition is given, only relations between them. Giving up the quest for an analysis of the nature of geometrical objects in profit of the axiomatic method has been a considerable scientific step. However, we contest such an analogy. Random objects are heavily used in many areas of science and technology: sampling, cryptology,... Of course, such objects are in fact "*as much as we can random*". Which means *fake randomness*. But they refer to an ideal notion of randomness which cannot be simply disregarded.

In fact, since Pierre Simon de Laplace (1749–1827), some probabilists never gave up the idea of formalizing the notion of random object. Let us cite particularly Richard von Mises (1883–1953) and Kolmogorov. In fact, it is quite impressive that, having so brilliantly and efficiently axiomatized probability theory via measure theory (Kolmogorov 1933), Kolmogorov was not fully satisfied of such foundations.[8] And he kept a keen interest to the quest for a formal notion of randomness initiated by von Mises in the 1920s.

3.4.1.2 The 100 Heads Paradox in Probability Theory

That probability theory fails to completely account for randomness is strongly witnessed by the following paradoxical fact. In probability theory, *if we toss an unbiaised coin 100 times then 100 heads are just as probable as any other outcome!* Who really believes that?

> The axioms of probability theory, as developped by Kolmogorov, do not solve all mysteries that they are sometimes supposed to.
>
> (Gács 1993)

3.4.1.3 Cryptology

Contrarily to probability theory, cryptology heavily uses random objects. Though again, no formal definition is given, random sequences are produced which are not fully random, just hard enough so that the mechanism which produces them cannot be discovered in reasonable time.

> Anyone who considers arithmetical methods of producing random reals is, of course, in a state of sin. For, as has been pointed out several times, there is no such thing as a random number — there are only methods to produce random numbers, and a strict arithmetical procedure is of course not such a method.
>
> (Von Neumann 1951)

So, what is "true" randomness? Is there something like a degree of randomness? Presently, (fake) randomness only means to pass some statistical tests. One can ask for more.

3.4.2 Kolmogorov's Proposal: Incompressible Strings

We now assume that $\mathcal{O} = \{0, 1\}^*$, i.e., we restrict to words.

---

[8] Kolmogorov is one of the rare probabilists – up to now – not to believe that Kolmogorov's axioms for probability theory do constitute the last word about formalizing randomness...

3.4.2.1 Incompressibility with Kolmogorov Complexity

Though much work had been devoted to get *a mathematical theory of random objects*, notably by Von Mises (1919, 1939), none was satisfactory up to the 1960s when Kolmogorov based such a theory on Kolmogorov complexity, hence on computability theory.

In fact, the theory was independently[9] developed by Gregory J. Chaitin (b. 1947), who submitted two papers (Chaitin 1966, 1969) in 1965. The basic idea is as follows:

- *Larger is the Kolmogorov complexity of a text, more random it is,*
- *Larger is its information content, and more compressed is the text.*

Thus, a theory for measuring the information content is also a theory of randomness.

Recall that there exists c such that for all $x \in \{0,1\}^*$, $K(x) \leq |x| + c$ (Proposition 3.1.15). The reason being that there is a "stupid" program of length about $|x|$ which computes the word x by telling what are the successive letters of x. The intuition of incompressibility is as follows: x is incompressible if there no shorter way to get x.

Of course, we are not going to define absolute randomness for words. But a measure of randomness telling *how far from* $|x|$ *is* $K(x)$.

Definition 3.4.1 (Measure of incompressibility). A word x is c-incompressible if $K(x) \geq |x| - c$.

It is rather intuitive that most things are random. The next Proposition formalizes this idea.

Proposition 3.4.2. *For any n, the proportion of c-incompressible strings of length n is* $\geq 1 - 2^{-c}$.

Proof. At most $2^{n-c} - 1$ programs of length $< n - c$ and 2^n strings of length n. □

3.4.2.2 Incompressibility with Length Conditional Complexity

We observed in Sect. 3.1.2.3 that the entropy of a word of the form $000\ldots0$ is null. i.e., entropy did not considered the information conveyed by the length. Here, with incompressibility based on Kolmogorov complexity, we can also ignore the information content conveyed by the length by considering *incompressibility based on length conditional Kolmogorov complexity.*

Definition 3.4.3 (Measure of length conditional incompressibility). A word x is length conditional c-incompressible if $K(x \mid |x|) \geq |x| - c$.

The same simple counting argument yields the following Proposition.

---

[9] Li and Vitányi (1997), pp. 89–92, gives a detailed account of *when who did what.*

Proposition 3.4.4. *For all n, the proportion of length conditional c-incompressible strings of length n is $\geq 1 - 2^{-c}$.*

A priori length conditional incompressibility is stronger than mere incompressibility. However, the two notions of incompressibility are about the same ...up to a constant.

Proposition 3.4.5. *There exists d such that, for all $c \in \mathbb{N}$, $x \in \{0,1\}^*$,*

1. x is length conditional c-incompressible $\Rightarrow$ x is $(c+d)$-incompressible
2. x is c-incompressible $\Rightarrow$ x is length conditional $(2c+d)$-incompressible.

Proof. 1 is trivial. For 2, observe that there is d such that, for all x,

(*) $\quad K(x) \leq K(x \mid |x|) + 2K(|x| - K(x \mid |x|)) + d$

In fact, if $K = K_\varphi$ and $K(\ \mid\ ) = K_{\psi(\ \mid\ )}$, consider p, q such that

$$\begin{array}{rr} |x| - K(x \mid |x|) = \varphi(p) & \psi(q \mid |x|) = x \\ K(|x| - K(x \mid |x|)) = |p| & K(x \mid |x|) = |q| \end{array}$$

With p and q, hence with $\langle p, q\rangle$ (cf. Proposition 3.1.5), one can successively get

$$\begin{cases} |x| - K(x \mid |x|) & \text{this is } \varphi(p) \\ K(x \mid |x|) & \text{this is } q \\ |x| & \text{just sum the above quantities} \\ x & \text{this is } \psi(q \mid |x|) \end{cases}$$

Thus, $K(x) \leq |\langle p, q\rangle| + O(1)$. Applying Proposition 3.1.5, we get (*).
Using $K^{\mathbb{N}} \leq \log + c_1$ and $K^{\{0,1\}^*}(x) \geq |x| - c$ (Proposition 3.1.15), (*) yields

$$|x| - K(x \mid |x|) \leq 2\log(|x| - K(x \mid |x|)) + 2c_1 + c + d$$

Finally, observe that $z \leq 2\log z + k$ insures $z \leq \max(8, 2k)$. □

3.4.3 Incompressibility Is Randomness (Martin-Löf)

Now, if incompressibility is clearly a necessary condition for randomness, how do we argue that it is a sufficient condition? Contraposing the wanted implication, let us see that if a word fails some statistical test then it is not incompressible. We consider some spectacular failures of statistical tests.

Example 3.4.6. 1. *[Constant half length prefix]* For all n large enough, a string $0^n u$ with $|u| = n$ cannot be c-incompressible.
2. *[Palindromes]* Large enough palindromes cannot be c-incompressible.
3. *[0 and 1 not equidistributed]* For all $0 < \alpha < 1$, for all n large enough, a string of length n which has $\leq \alpha\frac{n}{2}$ zeros cannot be c-incompressible.

Proof. 1. Let c' be such that $K(x) \leq |x| + c'$. Observe that there exists c'' such that $K(0^n u) \leq K(u) + c''$ hence

$$K(0^n u) \leq n + c' + c'' \leq \frac{1}{2}|0^n u| + c' + c''$$

So that $K(0^n u) \geq |0^n u| - c$ is impossible for n large enough.

2. Same argument: There exists c'' such that, for any palindrome x,

$$K(x) \leq \frac{1}{2}|x| + c''$$

3. The proof follows the classical argument to get the law of large numbers (cf. Feller 1968). Let us do it for $\alpha = \frac{2}{3}$, so that $\frac{\alpha}{2} = \frac{1}{3}$. Let A_n be the set of strings of length n with $\leq \frac{n}{3}$ zeros. We estimate the number N of elements of A_n.

$$N = \sum_{i=0}^{i=\frac{n}{3}} \binom{n}{i} \leq (\frac{n}{3} + 1) \binom{n}{\frac{n}{3}} = (\frac{n}{3} + 1) \frac{n!}{\frac{n}{3}! \, \frac{2n}{3}!}$$

Use inequality $1 \leq e^{\frac{1}{12n}} \leq 1.1$ and Stirling's formula (1730),

$$\sqrt{2n\pi} \left(\frac{n}{e}\right)^n e^{\frac{1}{12n+1}} < n! < \sqrt{2n\pi} \left(\frac{n}{e}\right)^n e^{\frac{1}{12n}}$$

Observe that $1.1\ (\frac{n}{3} + 1) < n$ for $n \geq 2$. Therefore,

$$N < n \frac{\sqrt{2n\pi}\,\left(\frac{n}{e}\right)^n}{\sqrt{2\frac{n}{3}\pi}\,\left(\frac{\frac{n}{3}}{e}\right)^{\frac{n}{3}} \sqrt{2\frac{2n}{3}\pi}\,\left(\frac{\frac{2n}{3}}{e}\right)^{\frac{2n}{3}}} = \frac{3}{2}\sqrt{\frac{n}{\pi}} \left(\frac{3}{\sqrt[3]{4}}\right)^n$$

Using Proposition 3.1.16, for any element of A_n, we have

$$K(x \mid n) \leq \log(N) + d \leq n \log\left(\frac{3}{\sqrt[3]{4}}\right) + \frac{\log n}{2} + d$$

Since $\frac{27}{4} < 8$, we have $\frac{3}{\sqrt[3]{4}} < 2$ and $\log\left(\frac{3}{\sqrt[3]{4}}\right) < 1$. Hence, $n - c \leq n \log\left(\frac{3}{\sqrt[3]{4}}\right) + \frac{\log n}{2} + d$ is impossible for n large enough.

So that x cannot be c-incompressible. □

Let us give a common framework to the three above examples so as to get some flavor of what can be a statistical test. To do this, we follow the above proofs of compressibility.

Example 3.4.7. 1. *[Constant left half length prefix].* Let V_m be the set of strings with m zeros ahead. The sequence $V_0, V_1, \ldots$ is decreasing. The number of strings of length n in V_m is 0 if $m > n$ and 2^{n-m} if $m \leq n$. Thus, the proportion $\frac{\sharp\{x \mid |x|=n \ \wedge \ x \in V_m\}}{2^n}$ of length n words in V_m is 2^{-m}.

2. *[Palindromes].* Put in V_m all strings which have equal length m prefix and suffix. The sequence $V_0, V_1, \ldots$ is decreasing. The number of strings of length n in V_m is 0 if $m > \frac{n}{2}$ and 2^{n-2m} if $m \leq \frac{n}{2}$. Thus, the proportion of length n words which are in V_m is 2^{-2m}.
3. *[0 and 1 not equidistributed].* Put in V_m^α = all strings x such that the number of zeros is $\leq (\alpha + (1-\alpha)2^{-m})\frac{|x|}{2}$. The sequence $V_0, V_1, \ldots$ is decreasing. A computation analogous to that done in the proof of the law of large numbers shows that the proportion of length n words which are in V_m is $\leq 2^{-\gamma m}$ for some $\gamma > 0$ (independent of m).

Now, what about other statistical tests? But what is a statistical test? A convincing formalization has been developed by Martin-Löf. The intuition is that illustrated in Example 3.4.7 augmented of the following feature: each V_m is computably enumerable and so is the relation $\{(m, x) \mid x \in V_m\}$. A feature which is analogous to the partial computability assumption in the definition of Kolmogorov complexity.

Definition 3.4.8 (Abstract notion of statistical test, Martin-Löf 1966, 1971). A statistical test is a family of nested critical sets

$$\{0,1\}^* \supseteq V_0 \supseteq V_1 \supseteq V_2 \supseteq \ldots \supseteq V_m \supseteq \ldots$$

such that $\{(m, x) \mid x \in V_m\}$ is computably enumerable and the proportion $\frac{\sharp\{x \mid |x|=n \ \wedge \ x \in V_m\}}{2^n}$ of length n words which are in V_m is $\leq 2^{-m}$.

Intuition. The bound 2^{-m} is just a normalization. Any bound $b(n)$ such that $b : \mathbb{N} \to \mathbb{Q}$ which is computable, decreasing and with limit 0 could replace 2^{-m}. The significance of $x \in V_m$ is that the hypothesis x *is random* is rejected with significance level 2^{-m}.

Remark 3.4.9. Instead of sets V_m one can consider a function $\delta : \{0,1\}^* \to \mathbb{N}$ such that $\frac{\sharp\{x \mid |x|=n \ \wedge \ \delta(x) \geq m\}}{2^n} \leq 2^{-m}$ and δ is computable from below, i.e., $\{(m, x) \mid \delta(x) \geq m\}$ is recursively enumerable.

We have just argued on some examples that all statistical tests from practice are of the form stated by Definition 3.4.8. Now comes Martin-Löf fundamental result about statistical tests which is in the vein of the invariance theorem.

Theorem 3.4.10 (Martin-Löf 1965). *Up to a constant shift, there exists a largest statistical test $(U_m)_{m \in \mathbb{N}}$; in terms of functions, up to an additive constant, there exists a largest statistical test Δ :*

$$\forall (V_m)_{m \in \mathbb{N}} \ \exists c \ \forall m \quad V_{m+c} \subseteq U_m \quad ; \quad \forall \delta \ \exists c \ \forall x \quad \delta(x) < \Delta(x) + c$$

Proof. Consider $\Delta(x) = |x| - K(x \mid |x|) - 1$.

$\boxed{\Delta \text{ is a test.}}$ Clearly, $\{(m, x) \mid \Delta(x) \geq m\}$ is c.e. $\Delta(x) \geq m$ means $K(x \mid |x|) \leq |x| - m - 1$. So no more elements in $\{x \mid \Delta(x) \geq m \wedge |x| = n\}$ than programs of length $\leq n - m - 1$, which is $2^{n-m} - 1$.

$\boxed{\Delta \text{ is largest.}}$ x is determined by its rank in the set $V_{\delta(x)} = \{z \mid \delta(z) \geq \delta(x) \wedge |z| = |x|\}$. Since this set has $\leq 2^{n-\delta(x)}$ elements, the rank of x has a binary representation of length $\leq |x| - \delta(x)$. Add useless zeros ahead to get a word p with length $|x| - \delta(x)$.

With p we get $|x| - \delta(x)$. With $|x| - \delta(x)$ and $|x|$ we get $\delta(x)$ and construct $V_{\delta(x)}$. With p we get the rank of x in this set, hence we get x. Thus, $K(x \mid |x|) \leq |x| - \delta(x) + c$, i.e., $\delta(x) < \Delta(x) + c$. □

A corollary of the previous result is that, for words, incompressibility implies (hence is equivalent to) randomness.

Corollary 3.4.11 (Martin-Löf 1965). *Incompressibility passes all statistical tests. I.e., for all c, for all statistical test $(V_m)_m$, there exists d such that*

$$\forall x \ (x \text{ is } c\text{-incompressible} \Rightarrow x \notin V_{c+d})$$

Proof. Let x be length conditional c-incompressible. This means that $K(x \mid |x|) \geq |x| - c$. Hence $\Delta(x) = |x| - K(x \mid |x|) - 1 \leq c - 1$, which means that $x \notin U_c$. Let now $(V_m)_m$ be a statistical test. Then there is some d such that $V_{m+d} \subseteq U_m$ Therefore $x \notin V_{c+d}$. □

Remark 3.4.12. Observe that incompressibility is a *bottom-up* notion: we look at the value of $K(x)$ (or that of $K(x \mid |x|)$). On the opposite, passing statistical tests is a *top-down* notion. To pass all statistical tests amounts to an inclusion in an intersection: namely, an inclusion in

$$\bigcap_{(V_m)_m} \bigcup_c V_{m+c}$$

3.4.4 Shortest Programs Are Random Finite Strings

Observe that optimal programs to compute any object are examples of random strings. More precisely, the following result holds.

Proposition 3.4.13. *Let $\mathbb{O}$ be an elementary set (cf. Definition 3.1.9) and $U : \{0,1\}^* \to \{0,1\}^*$, $V : \{0,1\}^* \to \mathbb{O}$ be some fixed optimal functions. There exists a constant c such that, for all $a \in \mathbb{O}$, for all $p \in \{0,1\}^*$, if $V(p) = a$ and $K_V(a) = |p|$ then $K_U(p) \geq |p| - c$. In other words, for any $a \in \mathbb{O}$, if p is a shortest program which outputs a then p is c-random.*

Proof. Consider the function $V \circ U : \{0,1\}^* \to \mathcal{O}$. Using the invariance theorem, let c be such that $K_V \leq K_{V \circ U} + c$. Then, for every $q \in \{0,1\}^*$,

$$\begin{aligned} U(q) = p &\Rightarrow V \circ U(q) = a \\ &\Rightarrow |q| \geq K_{V \circ U}(a) \geq K_V(a) - c = |p| - c \end{aligned}$$

Which proves that $K_U(p) \geq |p| - c$. □

3.4.5 *Random Finite Strings and Complexity Lower Bounds*

Random finite strings (or rather c-incompressible strings) have been extensively used to prove lower bounds for computational complexity, cf. the pioneering paper by Wolfgang Paul (1979) (see also an account of the proof in our survey paper Ferbus-Zanda and Grigorieff 2004) and the work by Li and Vitányi (1997). The key idea is that a random string can be used as a worst possible input.

3.5 Formalization of Randomness: Infinite Objects

We shall stick to infinite sequences of zeros and ones: $\{0,1\}^{\mathbb{N}}$.

3.5.1 *Martin-Löf Top-Down Approach: Effective Topology*

3.5.1.1 The Naive Idea Badly Fails

The naive idea of a random element of $\{0,1\}^{\mathbb{N}}$ is that of a sequence α which is in no set of measure 0. Alas, α is always in the singleton set $\{\alpha\}$ which has measure 0!

3.5.1.2 Martin-Löf's Solution: Effectivize

Martin-Löf's solution to the above problem is to extend to infinite sequences what he did for finite objects (cf. Sect. 3.4.3): effectivize. Now, this means to consider the sole effective measure zero sets.

Let us develop a series of observations which leads to Martin-Löf's precise solution, i.e., what does mean effective for measure 0 sets.

To prove a probability law amounts to prove that a certain set X of sequences has probability one. To do this, one has to prove that the complement set $Y = \{0,1\}^{\mathbb{N}} \setminus X$ has probability zero. Now, in order to prove that $Y \subseteq \{0,1\}^{\mathbb{N}}$ has probability zero, basic measure theory tells us that one has to include Y in open sets with arbitrarily small probability. I.e., for each $n \in \mathbb{N}$ one must find an open set $U_n \supseteq Y$ which has probability $\leq \frac{1}{2^n}$.

If things were on the real line $\mathbb{R}$ we would say that U_n is a countable union of intervals with rational endpoints. Here, in $\{0,1\}^{\mathbb{N}}$, U_n is a countable union of sets of the form $u\{0,1\}^{\mathbb{N}}$ where u is a finite binary string and $u\{0,1\}^{\mathbb{N}}$ is the set of infinite sequences which extend u.

In order to prove that Y has probability zero, for each $n \in \mathbb{N}$ one must find a family $(u_{n,m})_{m\in\mathbb{N}}$ such that $Y \subseteq \bigcup_m u_{n,m}\{0,1\}^{\mathbb{N}}$ and $Proba(\bigcup_m u_{n,m}\{0,1\}^{\mathbb{N}}) \leq \frac{1}{2^n}$ for each $n \in \mathbb{N}$.

Now, Martin-Löf makes a crucial observation: mathematical probability laws which we consider necessarily have some effective character. And this effectiveness should reflect in the proof as follows: *the doubly indexed sequence $(u_{n,m})_{n,m\in\mathbb{N}}$ is computable.* Thus, the set $\bigcup_m u_{n,m}\{0,1\}^{\mathbb{N}}$ is a *c.e. open set* and $\bigcap_n \bigcup_m u_{n,m}\{0,1\}^{\mathbb{N}}$ is a countable intersection of a *computably enumerable family of open sets*.

Now comes the essential theorem, which is analogous to Theorem 3.4.10.

Definition 3.5.1 (Martin-Löf 1966). A constructively null G_δ set is any set of the form

$$\bigcap_n \bigcup_m u_{n,m}\{0,1\}^{\mathbb{N}}$$

where $Proba(\bigcup_m u_{n,m}\{0,1\}^{\mathbb{N}}) \leq \frac{1}{2^n}$ (hence the intersection set has probability zero) and the sequence $u_{n,m}$ is computably enumerable.

Theorem 3.5.2 (Martin-Löf 1966). *There exist a largest constructively null G_δ set.*

Let us insist that the theorem says *largest*, up to nothing, really largest relative to set inclusion.

Definition 3.5.3 (Martin-Löf 1966). A sequence $\alpha \in \{0,1\}^{\mathbb{N}}$ is Martin-Löf random if it belongs to no constructively null G_δ set (i.e., if it does not belongs to the largest one).

In particular, the family of random sequences, being the complement of a constructively null G_δ set, has probability 1. And the observation above Definition 3.5.1 insures that Martin-Löf random sequences satisfy all usual probabilities laws. Notice that *the last statement can be seen as an improvement of all usual probabilities laws: not only such laws are true with probability* 1 *but they are true for all sequences in the measure* 1 *set of Martin-Löf random sequences.*

3.5.2 The Bottom-Up Approach

3.5.2.1 The Naive Idea Badly Fails

Another natural naive idea to get randomness for sequences is to extend randomness from finite objects to infinite ones. The obvious proposal is to consider sequences $\alpha \in \{0,1\}^{\mathbb{N}}$ such that, for some c,

$$\forall n \quad K(\alpha \restriction n) \geq n - c \tag{3.1}$$

However, Martin-Löf proved that there is no such sequence.

Theorem 3.5.4 (Large oscillations, (Martin-Löf 1971)). *If $f : \mathbb{N} \to \mathbb{N}$ is computable and $\sum_{n \in \mathbb{N}} 2^{-f(n)} = +\infty$ then, for every $\alpha \in \{0,1\}^{\mathbb{N}}$, there are infinitely many k such that $K(\alpha \restriction k) \leq k - f(k) - O(1)$.*

Proof. Let us do the proof in the case $f(n) = \log n$ which is quite limpid (recall that the harmonic series $\frac{1}{n} = 2^{-\log n}$ has infinite sum).

Let k be any integer. The word $\alpha \restriction k$ prefixed with 1 is the binary representation of an integer n (we put 1 ahead of $\alpha \restriction k$ in order to avoid a first block of non significative zeros). We claim that $\alpha \restriction n$ can be recovered from $\alpha \restriction [k+1, n]$ only. In fact,

- $n - k$ is the length of $\alpha \restriction [k+1, n]$,
- $k = \lfloor \log n \rfloor + 1 = \lfloor \log(n-k) \rfloor + 1 + \varepsilon$ (where $\varepsilon \in \{0,1\}$) is known from $n - k$ and ε,
- $n = (n-k) + k$.
- $\alpha \restriction k$ is the binary representation of n.

The above analysis describes a computable map $f : \{0,1\}^* \times \{0,1\} \to \{0,1\}^*$ such that $\alpha \restriction n = f(\alpha \restriction [k+1, n], \varepsilon)$. Applying Proposition 3.1.15, point 3, we get

$$K(\alpha \restriction n) \leq K(\alpha \restriction [k+1, n]) + O(1) \leq n - k + O(1) = n - \log(n) + O(1)$$

3.5.2.2 Miller and Yu's Theorem

It took about 40 years to get a characterization of randomness via Kolmogorov complexity which completes Theorem 3.5.4 in a very pleasant and natural way.

Theorem 3.5.5 (Miller and Yu 2008). *Let $\mathcal{F}$ be the family of total computable functions $f : \mathbb{N} \to \mathbb{N}$ satisfying $\sum_{n \in \mathbb{N}} 2^{-f(n)} < +\infty$. The following conditions are equivalent:*

i. *The sequence $\alpha \in \{0,1\}^{\mathbb{N}}$ is Martin-Löf random*
ii. $\exists c \quad \forall k \quad K(\alpha \restriction k) \geq k - f(k) - c$ *for every* $f \in \mathcal{F}$
iii. $\exists c \quad \forall k \quad K(\alpha \restriction k) \geq k - H(k) - c$

Moreover, there exists a particular $g \in \mathcal{F}$ *such that one can add a fourth equivalent condition:*

iv. $\exists c \quad \forall k \quad K(\alpha \restriction k) \geq k - g(k) - c$

Recently, an elementary proof of this theorem was given by Bienvenu et al. (2008). Equivalence $i \Leftrightarrow iii$ is due to Gács (1980).

3.5.2.3 Variants of Kolmogorov Complexity and Randomness

Bottom-up characterization of random sequences have been obtained using Levin monotone complexity, Schnorr process complexity and prefix complexity (cf. Sects. 3.3.1–3.3.3).

Theorem 3.5.6. *The following conditions are equivalent:*

i. The sequence $\alpha \in \{0,1\}^{\mathbb{N}}$ *is Martin-Löf random*
ii. $\exists c \quad \forall k \quad |K^{mon}(\alpha \restriction k) - k| \leq c$
iii. $\exists c \quad \forall k \quad |S(\alpha \restriction k) - k| \leq c$
iv. $\exists c \quad \forall k \quad H(\alpha \restriction k) \geq k - c$

Equivalence $i \Leftrightarrow ii$ is due to Levin (Zvonkin and Levin 1970). Equivalence $i \Leftrightarrow iii$ is due to Schnorr (1971a). Equivalence $i \Leftrightarrow iv$ is due to Schnorr and Chaitin (1975).

3.5.3 Randomness: A Robust Notion but a Fragile Property

3.5.3.1 Randomness: A Robust Mathematical Notion

Besides the top-down definition of Martin-Löf randomness, we mentioned above diverse bottom-up characterizations via properties of the initial segments with respect to variants of Kolmogorov complexity. There are other top-down and bottom-up characterizations, we mention two of them in this section. This variety of characterizations shows that Martin-Löf randomness is a robust mathematical notion.

3.5.3.2 Randomness and Martingales

Recall that a martingale is a function $d : \{0,1\}^* \to \mathbb{R}^+$ such that

$$\forall u \quad d(u) = \frac{d(u0) + d(u1)}{2}$$

The intuition is that a player tries to predict the bits of a sequence $\alpha \in \{0,1\}^{\mathbb{N}}$ and bets some amount of money on the values of these bits. If his guess is correct he doubles his stake, else he looses it. Starting with a positive capital $d(\varepsilon)$ (where ε is the empty word), $d(\alpha \upharpoonright k)$ is his capital after the k first bits of α have been revealed. The martingale d wins on $\alpha \in \{0,1\}^{\mathbb{N}}$ if the capital of the player tends to $+\infty$.

The martingale d is computably approximable from below if the left cut of $d(u)$ is computably enumerable, uniformly in u (i.e., $\{(u,q) \in \{0,1\}^* \times \mathbb{Q} \mid q \leq d(u)\}$ is c.e.).

Theorem 3.5.7 (Schnorr 1971b). *A sequence $\alpha \in \{0,1\}^{\mathbb{N}}$ is Martin-Löf random if and only if no martingale computably approximable from below wins on α.*

3.5.3.3 Randomness and Compressors

Recently, Bienvenu and Merkle obtained quite remarkable characterizations of random sequences in the vein of Theorems 3.5.6 and 3.5.5 involving *computable* upper bounds of K and H.

Definition 3.5.8. A compressor is any partial computable $\Gamma : \{0,1\}^* \to \{0,1\}^*$ which is one-to-one and has computable domain. A compressor is said to be prefix-free if its range is prefix-free.

Proposition 3.5.9. *1. For any computable upper bound F of K (resp. H) there exists a compressor (resp. prefix-free compressor) Γ such that*

$$\exists c \quad \forall x \in \{0,1\}^* \quad |\Gamma(x)| \leq F(x) + c$$

2. If Γ is a compressor (resp. prefix-free compressor) then

$$\exists c \quad \forall x \in \{0,1\}^* \quad K(x) \leq |\Gamma(x)| + c \qquad (resp. H(x) \leq |\Gamma(x)| + c)$$

Now comes the surprising characterizations of randomness in terms of *computable functions*.

Theorem 3.5.10 (Bienvenu and Merkle 2007). *The following conditions are equivalent:*

i. The sequence $\alpha \in \{0,1\}^{\mathbb{N}}$ is Martin-Löf random
ii. For all prefix-free compressor $\Gamma : \{0,1\}^* \to \{0,1\}^*$,

$$\exists c \quad \forall k \quad |\Gamma(\alpha \upharpoonright k)| \geq k - c$$

iii. For all compressor Γ, $\exists c \quad \forall k \quad |\Gamma(\alpha \upharpoonright k)| \geq k - H(k) - c$

Moreover, there exists a particular prefix-free compressor Γ^* *and a particular compressor* $\Gamma^\#$ *such that one can add two more equivalent conditions:*

iv. $\exists c \quad \forall k \quad |\Gamma^*(\alpha \restriction k)| \geq k - c$
v. $\exists c \quad \forall k \quad |\Gamma^\#(\alpha \restriction k)| \geq k - |\Gamma^*(\alpha \restriction k)| - c$

3.5.3.4 Randomness: A Fragile Property

Though the notion of Martin-Löf randomness is robust, with a lot of equivalent definitions, as a property, it is quite fragile.

In fact, random sequences loose their random character under very simple computable transformation. For instance, even if $a_0a_1a_2\ldots$ is random, the sequence $0a_00a_10a_20\ldots$ IS NOT random since it fails the following Martin-Löf test:

$$\bigcap_{n\in\mathbb{N}}\{\alpha \mid \forall i < n \ \alpha(2i+1) = 0\}$$

Indeed, $\{\alpha \mid \forall i < n \ \alpha(2i+1) = 0\}$ has probability 2^{-n} and is an open subset of $\{0,1\}^{\mathbb{N}}$.

3.5.4 Randomness Is Not Chaos

In a series of papers (Moschovakis 1993, 1994, 1996), Joan Rand Moschovakis introduced a very convincing notion of chaotic sequence $\alpha \in \{0,1\}^{\mathbb{N}}$. It turns out that the set of such sequences has measure zero and is disjoint from Martin-Löf random sequences.

This stresses that *randomness is not chaos*. As mentioned in Sect. 3.5.1.2, random sequences obey laws, those of probability theory.

3.5.5 Oracular Randomness

3.5.5.1 Relativization

Replacing "computable" by "computable in some oracle", all the above theory relativizes in an obvious way, using oracular Kolmogorov complexity and the oracular variants. In particular, when the oracle is the halting problem, i.e. the computably enumerable set $\emptyset'$, the obtained randomness is called 2-randomness. When the oracle is the halting problem of partial $\emptyset'$-computable functions, i.e. the computably enumerable set $\emptyset''$, the obtained randomness is called 3-randomness. And so on. Of course, 2-randomness implies randomness (also called 1-randomness) and 3-randomness implies 2-randomness. And so on.

3.5.5.2 Kolmogorov Randomness and $\emptyset'$

A natural question following Theorem 3.5.4 is to look at the so-called *Kolmogorov random sequences* which satisfy $K(\alpha \upharpoonright k) \geq k - O(1)$ for infinitely many k's. This question got a very surprising answer involving 2-randomness.

Theorem 3.5.11 (Nies et al. 2005). *Let $\alpha \in \{0,1\}^{\mathbb{N}}$. There are infinitely many k such that, for a fixed c, $K(\alpha \upharpoonright k) \geq k - c$ (i.e., α is Kolmogorov random) if and only if α is 2-random.*

3.5.6 Randomness: A New Foundation for Probability Theory?

Now that there is a sound mathematical notion of randomness, is it possible/reasonable to use it as a new foundation for probability theory? Kolmogorov has been ambiguous on this question. In his first paper on the subject, see pp. 35–36 of Kolmogorov (1965), he briefly evoked that possibility:

> ... to consider the use of the [Algorithmic Information Theory] constructions in providing a new basis for Probability Theory.

However, later, see pp. 35–36 of Kolmogorov (1983), he separated both topics:

> there is no need whatsoever to change the established construction of the mathematical probability theory on the basis of the general theory of measure. I am not enclined to attribute the significance of necessary foundations of probability theory to the investigations [about Kolmogorov complexity] that I am now going to survey. But they are most interesting in themselves.

though stressing the role of his new theory of random objects for *mathematics as a whole* in Kolmogorov (1983), p. 39:

> The concepts of information theory as applied to infinite sequences give rise to very interesting investigations, which, without being indispensable as a basis of probability theory, can acquire a certain value in the investigation of the algorithmic side of mathematics as a whole.

References

Becher, V., Figueira, S., Nies, A., & Picchi, S. (2005). Program size complexity for possibly infinite computations. *Notre Dame Journal of Formal Logic, 46*(1), 51–64.

Becher, V., Figueira, S., Grigorieff, S., & Miller, J. (2006). Random reals and halting probabilities. *The Journal of Symbolic Logic, 71*(4), 1411–1430.

Bienvenu, L., & Merkle, W. (2007). Reconciling data compression and Kolmogorov complexity. In *ICALP 2007*, Wroclaw (LNCS, Vol. 4596, pp. 643–654).

Bienvenu, L., Merkle, W., & Shen, A. (2008). A simple proof of Miller-Yu theorem. *Fundamenta Informaticae, 83*(1–2), 21–24.

Bonfante, G., Kaczmarek, M., & Marion, J.-Y. (2006). On abstract computer virology: From a recursion-theoretic perspective. *Journal in Computer Virology, 1*(3–4), 45–54.
Calude, C., & Jürgensen, H. (2005). Is complexity a source of incompleteness? *Advances in Applied Mathematics, 35*, 1–15.
Chaitin, G. (1966). On the length of programs for computing finite binary sequences. *Journal of the ACM, 13*, 547–569.
Chaitin, G. (1969). On the length of programs for computing finite binary sequences: Statistical considerations. *Journal of the ACM, 16*, 145–159.
Chaitin, G. (1971). Computational complexity & Gödel incompleteness theorem. *ACM SIGACT News, 9*, 11–12.
Chaitin, G. (1974). Information theoretic limitations of formal systems. *Journal of the ACM, 21*, 403–424.
Chaitin, G. (1975). A theory of program size formally identical to information theory. *Journal of the ACM, 22*, 329–340.
Delahaye, J. P. (1999). *Information, complexité, hasard* (2nd ed.). Paris: Hermès.
Delahaye, J. P. (2006). *Complexités : Aux limites des mathématiques et de l'informatique*. Montreal: Belin-Pour la Science.
Durand, B., & Zvonkin, A. (2004). Complexité de Kolmogorov. In E. Charpentier, A. Lesne, & N. Nikolski (Eds.), *L'héritage de Kolmogorov en mathématiques* (pp. 269–287). Berlin: Springer.
Fallis, D. (1996). The source of Chaitin's incorrectness. *Philosophia Mathematica, 4*, 261–269.
Feller, W. (1968). *Introduction to probability theory and its applications* (Vol. 1, 3rd ed.). New York: Wiley.
Ferbus-Zanda, M., & Grigorieff, S. (2004). Is randomnes native to computer science? In G. Paun, G. Rozenberg, & A. Salomaa (Eds.), *Current trends in theoretical computer science* (pp. 141–179). Singapore: World Scientific.
Ferbus-Zanda, M., & Grigorieff, S. (2006). Kolmogorov complexity and set theoretical representations of integers. *Mathematical Logic Quarterly, 52*(4), 381–409.
Gács, P. (1974). On the symmetry of algorithmic information. *Soviet Mathematics Doklady, 15*, 1477–1480.
Gács, P. (1980). Exact expressions for some randomness tests. *Zeitschrift für Mathematische Logik und Grundlagen der Mathematik, 26*, 385–394.
Gács, P. (1993). *Lectures notes on descriptional complexity and randomness* (pp. 1–67). Boston: Boston University. http://cs-pub.bu.edu/faculty/gacs/Home.html.
Huffman, D. A. (1952). A method for construction of minimum-redundancy codes. *Proceedings IRE, 40*, 1098–1101.
Knuth, D. (1981). *The art of computer programming. Volume 2: Semi-numerical algorithms* (2nd ed.). Reading: Addison-Wesley.
Kolmogorov, A. N. (1933). *Grundbegriffe der Wahscheinlichkeitsrechnung*. Berlin: Springer [Foundations of the theory of probability, Chelsea, 1956].
Kolmogorov, A. N. (1963). On tables of random numbers. *Sankhya, The Indian Journal of Statistics, Series A, 25*, 369–376.
Kolmogorov, A. N. (1965). Three approaches to the quantitative definition of information. *Problems of Information Transmission, 1*(1), 1–7.
Kolmogorov, A. N. (1983). Combinatorial foundation of information theory and the calculus of probability. *Russian Mathematical Surveys, 38*(4), 29–40.
Lacombe, D. (1960). La théorie des fonctions récursives et ses applications. *Bulletin de la Société Mathématique de France, 88*, 393–468.
van Lambalgen, M. (1989). Algorithmic information theory. *The Journal of Symbolic Logic, 54*(4), 1389–1400.
Levin, L. (1973). On the notion of a random sequence. *Soviet Mathematics Doklady, 14*, 1413–1416.

Levin, L. (1974). Laws of information conservation (non-growth) and aspects of the foundation of probability theory. *Problems of Information Transmission, 10*(3), 206–210.

Li, M., & Vitányi, P. (1997). *An introduction to Kolmogorov complexity and its applications* (2nd ed.). New York: Springer.

Martin-Löf, P. (1966). The definition of random sequences. *Information and Control, 9*, 602–619.

Martin-Löf, P. (1971). Complexity of oscilations in infinite binary sequences. *Zeitschrift für Wahrscheinlichkeitstheorie und verwandte Gebiete, 19*, 225–230.

Miller, J., & Yu, L. (2008). On initial segment complexity and degrees of randomness. *Transactions of the American Mathematical Society, 360*, 3193–3210.

Von Mises, R. (1919). Grundlagen der wahrscheinlichkeitsrechnung. *Mathematische Zeitschrift, 5*, 52–99.

Von Mises, R. (1939). *Probability, statistics and truth*. New York: Macmillan. (Reprinted: Dover 1981)

Moschovakis, J. R. (1993). An intuitionistic theory of lawlike, choice and lawless sequences. In J. Oikkonen & J. Väänänen (Eds.), *Logic colloquium '90*, Helsinski (Lecture notes in logic, Vol. 2, pp. 191–209). Berlin: Springer.

Moschovakis, J. R. (1994). More about relatively lawless sequences. *The Journal of Symbolic Logic, 59*(3), 813–829.

Moschovakis, J. R. (1996). A classical view of the intuitionistic continuum. *Annals of Pure and Applied Logic, 81*, 9–24.

Von Neumann, J. (1951). Various techniques used in connection with random digits. In A. S. Householder, G. E. Forsythe, & H. H. Germond (Eds.), *Monte Carlo method* (National bureau of standards applied mathematics series, Vol. 12, pp. 36–38). Washington, D.C.: U.S. Government Printing Office.

Nies, A., Stephan, F., & Terwijn, S. A. (2005). Randomness, relativization and Turing degrees. *The Journal of Symbolic Logic, 70*(2), 515–535.

Paul, W. (1979). Kolmogorov's complexity and lower bounds. In L. Budach (Ed.), *Proceedings of 2nd international conference on fundamentals of computation theory*, Berlin/Wendisch-Rietz (pp. 325–334). Berlin: Akademie.

Raatikainen, P. (1998). On interpreting Chaitin's incompleteness theorem. *Journal of Philosophical Logic, 27*(6), 569–586.

Russell, B. (1908). Mathematical logic as based on the theory of types. *American Journal of Mathematics, 30*, 222–262. (Reprinted in *From Frege to Gödel: A source book in mathematical logic, 1879–1931*, pp. 150–182, J. van Heijenoort, Ed., 1967)

Schnorr, P. (1971a). A unified approach to the definition of random sequences. *Mathematical Systems Theory, 5*, 246–258.

Schnorr, P. (1971b). *Zufälligkeit und Wahrscheinlichkeit* (Lecture notes in mathematics, Vol. 218). Berlin/New York: Springer.

Schnorr, P. (1973). A process complexity and effective random tests. *Journal of Computer and System Sciences, 7*, 376–388.

Shannon, C. E. (1948). The mathematical theory of communication. *Bell System Technical Journal, 27*, 379–423.

Soare, R. (1996). Computability and recursion. *Bulletin of Symbolic Logic, 2*, 284–321.

Solomonov, R. (1964a). A formal theory of inductive inference, part I. *Information and Control, 7*, 1–22.

Solomonov, R. (1964b). A formal theory of inductive inference, part II. *Information and Control, 7*, 224–254.

Zvonkin, A., & Levin, L. (1970). The complexity of finite objects and the development of the concepts of information and randomness by means of the theory of algorithms. *Russian Mathematical Surveys, 6*, 83–124.

Chapter 4
Kolmogorov Complexity in Perspective Part II: Classification, Information Processing and Duality

Marie Ferbus-Zanda

Abstract We survey diverse approaches to the notion of information: from Shannon entropy to Kolmogorov complexity. Two of the main applications of Kolmogorov complexity are presented: randomness and classification. The survey is divided in two parts in the same volume.

Part II is dedicated to the relation between logic and information system, within the scope of Kolmogorov algorithmic information theory. We present a recent application of Kolmogorov complexity: classification using compression, an idea with provocative implementation by authors such as Bennett, Vitányi and Cilibrasi among others. This stresses how Kolmogorov complexity, besides being a foundation to randomness, is also related to classification. Another approach to classification is also considered: the so-called "Google classification". It uses another original and attractive idea which is connected to the classification using compression and to Kolmogorov complexity from a conceptual point of view. We present and unify these different approaches to classification in terms of Bottom-Up versus Top-Down operational modes, of which we point the fundamental principles and the underlying duality. We look at the way these two dual modes are used in different approaches to information system, particularly the relational model for database introduced by Codd in the 1970s. These operational modes are also reinterpreted in the context of the comprehension schema of axiomatic set theory ZF. This leads us to develop how Kolmogorov's complexity is linked to intensionality, abstraction, classification and information system.

M. Ferbus-Zanda (✉)
LIAFA, CNRS & Université Paris Diderot – Paris 7, Case 7014, 75205 Paris Cedex 13, France
e-mail: Marie.Ferbus@liafa.univ-paris-diderot.fr

J. Dubucs, M. Bourdeau (eds.), *Constructivity and Computability in Historical and Philosophical Perspective*, Logic, Epistemology, and the Unity of Science 34,
DOI 10.1007/978-94-017-9217-2_4

Note. All notations and definitions relative to Kolmogorov complexity are introduced in Part I.[1]

4.1 Algorithmic Information Theory and Classification

Using Andrei Nikolaevich Kolmogorov complexity, striking results have been obtained on the problem of classification for quite diverse families of objects: let them be literary texts, music pieces, examination scripts (lax supervised) or, at a different level, natural languages and natural species (phylogeny).

The authors, mainly Charles Bennett, Paul Vitányi, Rudi Cilibrasi[2] have worked out refined methods which are along the following lines.

4.1.1 Definition and Representation of the Family of Objects We Want to Classify

First we have to define a specific family of objects which we want to classify. For example, a set of Russian literary texts that we want to group by authors. In this simple case, all texts are written in their original Russian language. Another instance, music. In that case, a common translation is required, i.e., a *normalization* of music pieces (representing or, in other words, interpreting musical partitions) which we want to group by composer. This common representation (which has to be tailored for computer programs) is necessary in order to be able to compare these diverse music pieces. Let us cite Delahaye (2004): Researchers considered 36 music pieces coded as MIDI (*Musical Instrumental Digital Interface*) files. They normalized them by producing piano versions and considering them as data files consisting of long lists of bytes.[3] Without such a normalization, which is a real informations extraction, nothing would work [...]. An instance at a different level: the 52 main European languages. In that case one has to choose a canonical object (here a text) and its representations (here translations) in each one of the different languages (i.e. corpus) that we consider. For instance, the *Universal Declaration of Human Rights* and its translations in these languages, an example which was a basic

[1]One can also consult Ferbus-Zanda and Grigorieff (2001), Durand and Zvonkin (2004), Delahaye (1999), Li and Vitányi (1997) and the pioneer works Kolmogorov (1965), Chaitin (1966, 1975) and Solomonoff (1964).

[2]One can read the surveys (Delahaye 2004, 2006) that give a clear introduction to these works (let us acknowledge that they were very helpful for us).

[3]A *byte* is a sequence of eight binary digits. It can also be seen as a number between 0 and 255.

test for Vitányi's method. As concerns natural species (another example developed by Vitányi), the canonical object will be a DNA sequence.

What has to be done is to select, define and normalize a family of objects or a corpus that we want to classify.

Normalization of a family of objects is a complex problem, and it may be also the case for the definition of such a family. *Roughly speaking*, one can partition the types of considered objects in different classes:

- Well defined families of objects to be classified. Normalization of these objects (rather of their representations) can then be done without loss of information. This is the case of literary texts.
- The family to be classified can be finite though unknown, possibly without a priori bound on its size. Such is the case with informations on the Web (cf. classification using Google, Sect. 4.3).
- There are some cases where such a normalization is difficult to work out if not impossible. It may be the case for painting, drawing, photography, art-house cinema, etc.

4.1.2 Comparing the Common Information Content

Finally, one gets a family of words in the same alphabet which represent the objects that we want to compare and measure the common information content[4] (observe that we can reduce to a binary alphabet). Our goal is to compare and, if possible, to measure the common information content. This is done by defining a distance for the pairs of such (binary) words with the following intuition: *The more common information is shared by two words, the closer they are and the shorter is their distance. Conversely, the less common information existing between two words, the more they are independent and non correlated, and greater is their distance. Two identical words have a null distance. Two totally independent words (for example, words representing two events as* 100 *random coin tosses) have a distance of about* 1 *(for a normalized distance bounded by* 1*)*. Observe that the authors, in their approach of classification of information, follow the ideas pioneered by Claude Shannon and Kolmogorov to define a *quantitative* measure of information content of words, i.e. a measure of their randomness (in exactly the same way as a volume or a surface gets a numerical measure).

[4]The notion of information content of an object is detailed in Part I. According to Kolmogorov, this is, by definition, the algorithmic complexity of that object.

4.1.3 Classification

We now have to associate a classification to the objects or corpus defined in Sect. 4.1.1 using the numerical measures based on the distances introduced in Sect. 4.1.2. This step is presently the least formally defined. The authors give representations of the obtained classifications using tables, trees, graphs, etc. This is indeed more a *visualization*, i.e. a *graphic representation*, of the obtained classification than a *formal classification*. Here the authors have no powerful mathematical framework such as the *relational model for databases* elaborated by Edgar F. Codd in the 1970s (Codd 1970) and its (recent) extension to *object database* with trees. Codd's approach is currently one of the sole mathematical formal approaches (if not the only one) to the notion of *information structuralization*. In this way, one can say that structuralization a class of informations or (representations of) objects (from the "real world" as computer scientists call it) amounts to a relational database which is itself a perfectly defined mathematical object. Moreover one can question this database and extract "new" informations via *queries* which can be written in a formal language (namely *Codd's relational algebra*). Also, notice that this extremely original theoretical approach is the one which is implemented in all database softwares since the 1980s and is used everywhere there is some mention of databases.

Consequently, the question is how are we to interpret in a formal way tables or trees in classification via compression and more particularly how are we to formally extract informations from this classification? Though valuable, the classification obtained by this method (of classification via compression) is rudimentary and *non formal*. This is somewhat analogous, for instance, to the classification of words in a dictionary of synonyms. We face a complex problem on which we shall return in Sect. 4.4. Nevertheless, Vitányi et al. obtained by these methods a classification tree for the 52 European languages which is the one revealed by linguists, a remarkable success. They also obtained phylogenetic trees classifying natural species which are in accordance with those obtained by paleontologists. These trees represent parenthood relations between natural species and are obtained via DNA sequence comparisons.

4.1.4 Normalization

An important problem remains when using a distance as in Sect. 4.1.3. To obtain a classification, we have to consider the amount of information contained in the considered objects. Let us cite Cilibrasi (2003): Large objects (in the sense of long strings) that differ by a tiny part are intuitively closer than tiny objects that differ by the same amount. For example, two whole mitochondrial genomes of 18,000 bases that differ by 9,000 are very different, while two whole nuclear genomes of 3×10^9 bases that differ by only 9,000 bases are very similar. As we shall see later, this problem is relatively easy to solve using a normalization of distances. Notice that this is a different way of normalization that the one proposed in Sect. 4.1.1.

4.1.5 Compression

Finally, all these methods rely on Kolmogorov complexity which is, as we know, a non computable function (cf. for example Ferbus-Zanda and Grigorieff 2001). The remarkable idea introduced by Vitányi is the following:

- *Consider the Kolmogorov complexity of an object as the ultimate, ideal and optimal value of the compression of the representation of that object.*
- *Compute approximations of this ideal compression using usual efficient compression algorithms, implemented with compressors such as gzip, bzip2, PPM, etc. which are of common use with computers.*

Observe that the quality and fastness of such compressors is largely due to heavy use of statistical tools. For example, *PPM* (*Prediction by Partial Matching*) uses a pleasing mix of statistical[5] models arranged by trees, suffix trees or suffix arrays.

We remark that the efficiency of these tools is of course due to several dozens of years of research in data compression. And as time goes on, they improve and better approximate Kolmogorov complexity. Replacing the "pure" but non computable Kolmogorov complexity by a banal compression algorithm such as gzip is quite a daring step taken by Vitányi.

4.2 Classification via Compression

4.2.1 The Normalized Information Distance (NID)

We now formalize the notions described above. The basic idea is to measure the information content shared by two binary words representing some objects in a family we want to classify.

The first such tentative goes back to the 1990s (Bennett et al. 1998): Bennett et al. define a notion of *information distance* between two words x, y as the size of the shortest program which maps x to y and y to x. This notion relies on the idea of *reversible computation*. A possible formal definition for such a distance is:

$$ID'(x, y) = \text{least } |p| \text{ such that } U(p, x) = y \text{ and } U(p, y) = x$$

where $U : \{0,1\}^* \times \{0,1\}^* \to \{0,1\}^*$ is optimal for the conditional complexity $K(\ |\ )$ (cf. Part I).

We shall mainly work with the following alternative definition:

$$ID(x, y) = \max\{K(x|y), K(y|x)\}$$

[5] We come back in Sect. 4.4 below on information processing with statistics.

The intuition for these definitions is that the shortest program which computes x from y and y from x takes into account all similarities between x and y. Observe that the two definitions do not coincide (even up to logarithmic terms) but lead to similar developments and efficient applications.

Note. In the definition of *ID*, we can consider K to be plain Kolmogorov complexity or its prefix version (denoted H below). In fact, this does not matter for a simple reason: all properties involving this distance will be true up to a $O(\log(|x|), \log(|y|))$ term and the difference between $K(z|t)$ and $H(z|t)$ is bounded by $2\log(|z|)$. For conceptual simplicity, we stick to plain Kolmogorov complexity. *ID* and ID' satisfy the axioms of a distance *up to a logarithmic term*.

The strict axioms for a distance d are

$$\begin{cases} d(x,x) = 0 & \textit{(identity)} \\ d(x,y) = d(y,x) & \textit{(symmetry)} \\ d(x,z) \le d(x,y) + d(y,z) & \textit{(triangle inequality)} \end{cases}$$

Theorem. *The up to a* log *term distance axioms which are satisfied by* ID *and* ID′ *are as follows:*

$$\begin{cases} d(x,x) = O(1) & (1) \\ d(x,y) = d(y,x) & (2) \\ d(x,z) \le d(x,y) + d(y,z) + O(\log(d(x,y) + d(y,z))) & (3) \end{cases}$$

Proof. We only treat the case of *ID*. Let $f : \{0,1\}^* \times \{0,1\}^* \to \{0,1\}^*$ be such that $f(p,x) = x$ for all p, x. The invariance theorem insures that $K(x|x) \le K_f(x|x) + O(1)$. Now, taking p to be the empty word, we see that $K_f(x|x) = 0$. Thus, $ID(x,x) = O(1)$.

Equality $ID(x,y) = ID(y,x)$ is obvious.

Let now p, p', q, q' be shortest programs such that $U(p,y) = x$, $U(p',x) = y$, $U(q,z) = y$, $U(q',y) = z$. Thus, $K(x|y) = |p|$, $K(y|x) = |p'|$, $K(y|z) = |q|$, $K(z|y) = |q'|$. Consider the injective computable function $\langle\,\rangle : \{0,1\}^* \times \{0,1\}^* \to \{0,1\}^*$ (cf. Proposition 1.6 in Part I) which is such that $|\langle r,s\rangle| = |r| + |s| + O(\log|r|)$. Let $\varphi : \{0,1\}^* \times \{0,1\}^* \to \{0,1\}^*$ be such that $\varphi(\langle r,s\rangle, x) = U(s, U(r,x))$. Then

$$\begin{aligned} \varphi(\langle q,p\rangle, z) &= U(p, U(q,z)) &= U(p,y) &= x \\ \varphi(\langle p',q'\rangle, x) &= U(q', U(p',x)) &= U(q',y) &= z \end{aligned}$$

so that, applying the invariance theorem, we get

$$\begin{aligned} K(x|z) &\le K_\varphi(x|z) + O(1) \le |\langle q,p\rangle| + O(1) \\ &= |q| + |p| + O(\log(|q|)) = K(y|z) + K(x|y) + O(\log(K(y|z))) \end{aligned}$$

and, similarly, $K(z|x) \leq K(y|x) + K(z|y) + O(\log(K(z|y)))$. Thus,

$$\begin{aligned}\max(K(x|z), K(z|x)) &\leq \max(K(y|z) + K(x|y) + O(\log(K(y|z))),\\ &\qquad K(y|x) + K(z|y) + O(\log(K(z|y))))\\ &\leq \max(K(x|y), K(y|x)) + \max(K(y|z), K(z|y))\\ &\qquad + O(\log(\max(K(y|z), K(z|y))))\end{aligned}$$

Which means $ID(x,z) \leq ID(x,y) + ID(y,z) + O(\log(ID(y,z)))$, a slightly stronger result than (3). □

It turns out that such approximations of the axioms are enough for the development of the theory.

To avoid scale distortion, as said in Sect. 4.1.4, distance *ID* is normalized to *NID* (*normalized information distance*) as follows, Li et al. (2003):

$$NID(x,y) = \frac{ID(x,y)}{\max(K(x), K(y))}$$

The remaining problem is that this distance is not computable since K is not. Here comes Vitányi's daring idea: *Consider* NID *as an ideal distance which is to be approximated by replacing the Kolmogorov function K by computable approximations obtained via compression algorithms.*

4.2.2 *The Normalized Compression Distance* (NCD)

The approximation of $K(x)$ by $\Gamma(x)$ where Γ is a compressor,[6] does not suffice. We also have to approximate the conditional Kolmogorov complexity $K(x|y)$. Vitányi chooses the following approximation:

$$\Gamma(y|x) = \Gamma(xy) - \Gamma(x)$$

The authors explain as follows their intuition: *To compress the word xy (x concatenated to y)*

- *The compressor first compresses x.*
- *Then it compresses y but skips all information from y which was already in x.*

Thus, the output is not a compression of y but a compression of y with all x information removed, i.e. this output is a conditional compression *of y knowing x.*

---

[6]A formal definition of compressors is given in Part I.

Now, the assumption that in the compression of the word xy the compressor first compresses x is questionable: how does the compressor recovers x in xy? One can argue positively in trivial case x and y are random (i.e. incompressible) and in case $x = y$. And between these two extreme cases? But it works. The miracle of modeling? Or something not completely understood?

With this approximation, plus the assumption that $\Gamma(xy) = \Gamma(yx)$ (also questionable: it depends on the used compressor) we get the following approximation of *NID*, called the *normalized compression distance*, *NCD*:

$$\begin{aligned} NCD(x, y) &= \frac{\max\,(\Gamma(x|y),\ \Gamma(y|x))}{\max\,(\Gamma(x),\ \Gamma(y))} \\ &= \frac{\max\,(\Gamma(yx) - \Gamma(y),\ \Gamma(xy) - \Gamma(x))}{\max\,(\Gamma(x),\ \Gamma(y))} \\ &= \frac{\Gamma(xy) - \min\,(\Gamma(x),\ \Gamma(y))}{\max\,(\Gamma(x),\ \Gamma(y))} \end{aligned}$$

Remark that clustering according to *NCD* and, more generally, classification via compression, is a *black box*[7] as noticed by Delahaye (2006): words are grouped together according to features that are not explicitly known to us, except if we had already a previous idea. Moreover, there is no reasonable hope that the analysis of the computation done by the compressor gives some light on the obtained clusters. For example, what makes a text by Tolstoi so characteristic? What differentiates the *styles* of Tolstoi and Dostoievski? But it works, Russian texts are grouped by authors by a compressor which ignores everything about Russian literature.... When dealing with some classification obtained by compression, one should have some idea about this classification: this is semantics whereas the compressor is purely syntactical and does not "understand" anything". Thus one cannot hope some help in understanding (interpretation) of the obtained classification (cf. Sect. 4.4). This is very much as with machines which, given some formal deduction system, are able to prove quite complex statements. But these theorems are proved with no explicit semantical idea, how are we to interpret them? No hope that the machine gives any hint, at least in the present context.

[7]The notion of black box is a scientific concept introduced by Norbert Wiener (1948). This concept is one of the fundamental principles of Cybernetics. It is issued from the multidisciplinary exchanges during the Macy conferences which were held in New-York, 1942–1953. Indeed, the emergence of cybernetics and information theory owes much to these conferences.

4.3 The Google Classification

Though *stricto sensu*, it does not use Kolmogorov complexity, we now present another recent approach (Cilibrasi and Vitányi 2007) to classification which leads to a very performing tool.

4.3.1 The Normalized Google Distance (NGD)

This quite original method is based on the huge data mass, constituted by the Web and which is accessible with search engines as Google. They allow for basic queries using a simple keyword or conjunction of keywords. Observe that the Web (the *World Wide Web*) is not a formal database: it is merely a crude data bank, in fact a (gigantic) informal information system since data on the Web are not structured as data in relational database. It has a rudimentary form of structuralization based on graphs and *graphical user interfaces*. Nevertheless, it is endowed with an object-oriented programming language, namely, Java. What is remarkable is that there exists a norm for this programming language and, moreover, this language is Turing-complete (cf. Sect. 4.4.2). This can explain the success (and fashion) of Java and of the *object approach* which is also largely due to the success of the Web.

Citing Evangelista and Kjos-Hanssen (2006), the idea of the method is as follows: When the Google search engine is used to search for the word x, Google displays the number of hits that word x has. The ratio of this number to the total number of Web pages indexed by Google represents the probability that word x appears on a Web page [...] If word y has a higher conditional probability to appear on a Web page, given that word x also appears on that Web page, than it does by itself, then it can be concluded that words x and y are related. Let us cite an example Cilibrasi and Vitányi (2005), which we complete with updated figures.[8] The searches for the index term "horse", "rider" and "molecule" respectively return 156, 62.2 and 45.6 million hits. Searches for pairs of words "horse rider" and "horse molecule" respectively return 2.66 and 1.52 million hits. These figures stress a stronger relation between the words "horse" and "rider" than between "horse" and "molecule".

Another example with famous paintings: "Le déjeuner sur l'Herbe", "Le Moulin de la Galette" and "La Joconde". Let refer them by a, b, c. Google searches for a, b, c respectively give 446,000, 278,000 and 1,310,000 hits. As for the searches for the conjunctions a + b, a + c and b + c, they respectively give 13,700, 888 and 603 hits. Clearly, Jean Renoir's paintings are more often cited together than each one is with Leonardo da Vinci's paintings.

In this way, the method regroups paintings by artists, using what is said about these paintings on the Web. But this does not associate the painters to groups of

[8] Point 4, Sect. 4.3.2 relativizes the obtained results.

paintings (we have to add them "by hand"). Formally, one can define the *normalized Google distance* as follows (Cilibrasi and Vitányi 2005, 2007):

$$NGD(x, y) = \frac{\max(\log \Lambda(x), \log \Lambda(y)) - \log \Lambda(x, y)}{\log \Upsilon - \min(\log \Lambda(x), \log \Lambda(y))}$$

where $\Lambda(z_1, \dots z_n)$ is the number of hits for the conjunctive query $z_1 \wedge \dots \wedge z_n$ (which is $z_1 \dots z_n$ with Google; If $n = 1$, $\Lambda(z)$ is the total number of hits for the query z). Υ is the total number of Web pages that Google indexes.

4.3.2 Discussing the Method

Let us cite some points relative to the use of such a classification method (the list is not exhaustive):

1. The number of objects in a future classification and that of canonical representatives of the different corpora is not chosen in advance nor even boundable in advance and it is constantly moving. This dynamical and uncontrolled feature of a definition of a family is a totally new experience, at least for a formal approach of classification.
2. Domains a priori completely rebel to classification such as the pictorial domain (a priori no normalization of paintings being possible or if it is this is not obvious in the present context) can now be easily considered. This is also the case (and for the same reasons) for sculpture, architecture, photography, art-house cinema, etc. This is so because we are no more dealing with the works themselves but with a discourse about them (which is the one on the Web). This speech depends on a "real" language: a natural language or a formal language. Notice that the notion of "pictorial language" remains a metaphor as long as we consider that infra verbal communication is not a language in the usual sense. *The discourse which is considered by Google is the one of the keywords and relations between them, these keywords coming from queries proposed for the* NGD *and appearing in the texts of the users of the Web.* Notice that here are some works which can be used for an algorithmic approach (possibly a normalization) of pictural pieces, art-house films, etc. For instance, the French psychoanalyst Murielle Gagnebin elaborated a theory of *esthetics* and *creation*, based on psychoanalysis and philosophy. Her meta psychological model is quite efficient to point out the fundamental psychical mechanisms involved in art pieces. And this is done from the art pieces themselves, excluding any discursive consideration on these art pieces or on the artists. Such a model could much probably be implemented as an expert system.
3. However, there is a big limitation to the method, namely that one which is called: *the closed world assumption*. That can be interpreted as follow: *the world*

according to Google,[9] *information according to Google*, etc. If Google finds something, how can one check its *pertinence*. Else, what does it mean? How can we define (in a general manner) a notion of pertinence for the informations found by Google? Sole certainty, that of uncertainty! Moreover, we notice that when failing to get hits with several keywords, we give up the original query and modify (we change its semantics) it up to the point Google gives some "pertinent" answers. That sort of failure is similar to the use of negation in the Prolog programming language (called *negation as failure*), which is much weaker than classical negation and which is connected to the closed world assumption for databases.

When failing to get hits, it is reasonable to give up the query and accordingly consider the related conjunction as meaningless. However, one should keep in mind that this is relative to the closed, and relatively small, world of data on the Web, the sole world accessible to Google. Also one has not to underestimate the changing aspect of the informations available on the Web. When succeeding with a query, the risk is to stop on this succeeding query and

- Forget that previous queries have been tried and have failed.
- Omit going on with some other queries which could possibly lead to more pertinent answers.
- Given a query, the answers obtained from Google are those found at a given moment in a kind of *snapshot* of the Web. But such an instantaneous snapshot betrays what is the essence of the Web: to be a continuously moving information system. All the updates (insertions, deletions, corrections, etc.) are done in a massively parallel context since Google uses about 700,000 computers as servers! Thus, Google answers to a query are not at all *final* answers nor do they constitute *an absolute answer*. And this is in contrast with the perfect determinism we are used when computer programs are run (in this way, Prolog is considerably more "deterministic" than Google) or with databases (when they are well written...) Also, the diverse answers given by Google may contradict one another, depending on the sites Google retained. In particular, one is tempted to stop when a site is found that gives an answer which seems convenient (indeed, this is what we do in most cases).

4. So we see some difficulties emerging with the theoretical approach to how Google deals with information from the Web (and the same for any browser). For such a reflection, we have chosen an idealistic perspective where Google searches according to scientific criteria or at least with some transparency (in particular, on how Web pages are indexed, or even how many are really indexed). However let us mention that there are some controversies about the indexing and consequently on exactness of the results found by Google, in particular, about the number of occurrences of a given word of all existing Web pages (even if not dealing with the content of these pages). Indeed, some queries lead to very surprising results:

[9]Irving (1978).

"Googlean logic" is quite strange (when compared with Boolean logic). This is shown in a very striking (and completely scientific) manner by Jean Véronis in his blog.[10]

A highly important task remains to be done in order to formalize the notion of information on the Web and the relations ruling the data it contains, as it has been done by Codd with the relational model for databases in the 1970s. Previous to Codd's work, organizing and structuralization data and information in a computer and their accessibility via the notion of query was underlaid by no solid mathematical foundation and was resting on technical tricks. This is still the case for the data on the Web. This remarkable innovative approach via Google search is still in its infancy.

In the next sections, we consider some formalized notions together with not yet formalized ideas – such as those pointed out in Sect. 4.1.3 – Ongoing work in progress and some papers are in preparation (Ferbus-Zanda In preparation-c, In preparation-d).

4.4 Classification, Bottom-Up Versus Top-Down Approaches and Duality

4.4.1 Bottom-Up Versus Top-Down Modes

These approaches to classification via compression and Google search (relative to information appearing on the Web) are incredibly original and present a huge interest. With the phenomenal expansion of computer science, nets and the Web, information has a kind of new status. So that these approaches (which are indeed based on what they are able to make explicit) help us to grasp this entirely new status of information as it is now with such a world of machines. Whereas classification via the relational model for databases has a neat formalization, we have stressed above how difficult it is to formally define the classification obtained by compression or via Google. Of course, one could base such a formalization on trees and graphs. But with such structures, the way information is recovered is rather poorly formalized. This is in fact what happens with the organization of files in an *operating system* since none of them uses any database (let it be *Unix, Linux, MacOs, Windows* and their variants).

It seems to us that one should reconsider these different approaches to the notion of classification in terms of *two fundamental modes to define mathematical and computer science objects which are also found in the execution of computer programs*. These two main approaches to define mathematical and computer science objects are:

[10] Véronis (2005).

- *Iterative definitions* (based on *set theoretical union*)
- *Inductive (or recursive) definitions* (based on *set theoretical intersection*).

For instance, one can define propositional formulas, terms and first-order logic formulas following any one of these two ways.

Recall that Stephen Kleene's presentation[11] of partial recursive functions is based on three (meta) operations: *composition*, *primitive recursion* and *minimization*.

- Iterative definitions are connected to minimization (and to the notion of *successor*). We can describe these type of definitions as "*Bottom-Up*" characterizations.
- Inductive definitions are connected to primitive recursion (and to the notion of *predecessor*). We can describe these type of definitions as "*Top-Down*" characterizations.

Notice that composition is related to both characterizations, the bottom-up and top-down ones. We gave, in Part I, formalizations of randomness for infinite objects which follow these two bottom-up and top-down approaches (cf. Part I, Sects. 5.1 and 5.2).

These two modes are also found in the execution of computer programs:

- *Execution in the iterative mode* is called *Bottom-Up*.
- *Execution in the recursive mode* is called *Top-Down*.

This last mode requires the use of a *stack* which goes on growing and decreasing and into which results of intermediate computations are pushed until getting to the "basic cases", i.e. the initial steps of the inductive definition of the program which is executed. To execute an iterative program, all data necessary for its execution are at disposal without need of any stack. From the computer scientist point of view, these two execution modes are really far apart. Notice that the execution mode (iterative or recursive) follows the definition mode (iterative or recursive) of the program to be executed. Nevertheless, in some cases, recursive programs may be executed in an iterative way avoiding any stack.[12]

In the same way, one observes that there are *two modes* – let us also call them *Bottom-Up* and *Top-Down* – that are used in the approach to classification of information and/or objects (of the real world) which are formally represented as

---

[11] Kleene formally and completely characterizes the notion of *recursive function* (also called *computable function*), by adding the minimization schema (1936) to the composition and recursion schemas – these two last schemas characterize the *primitive recursive functions* which constitute a proper subclass of the class of computable functions: the *Ackermann function* (1928) is computable but not primitive recursive. From a programming point of view, the minimization schema corresponds to the `while` loop (`while F(x) do P` where `F(x)` is a Boolean valued property and `P` is a program) (cf. the book by Shoenfield (2001)).

[12] This is the case for *tail-recursion* definitions. In some programming languages such as *LISP* such tail-recursion programs are generally executed (when the programs executor is well written) in an iterative way. Tail-recursion programs represent a *limit case* between iterative programs and recursive programs.

words or more generally as texts[13] or even as sets of words, in some alphabet (which can, as usual, be supposed to be binary).

- *In the Bottom-Up mode, one enters into information details. Otherwise said, one accesses the content of texts, i.e. the words representing the diverse informations and/or objects that one wants to classify (and the meaning of these words and/or texts). Texts, families of words, etc. are grasped from the* inside *and their meaning is essential.*
- *In the Top-Down mode, one does not access the content of texts in the above way. Texts are, in fact, handled from the* outside, *that is "from the top and down".*[14] *To say things otherwise, one uses a kind of "oracle" to grasp texts and families of words, i.e. means that are* exterior *to the understanding of text and the content of words.* Let us illustrate this with an example: *the use of keywords to structure families of texts.* One then uses both bottom-up and top-down modes to classify texts in the following way:

1. It is usual to follow a bottom-up approach in the choice of keywords. Particular words in texts are *chosen in consideration of the content of texts and their meaning* and in order to facilitate future searches. More precisely, some words will be considered as keywords and will be so declared. This is typically the case with scientific papers where keywords are chosen by the author, the journal editor, the librarian, etc. in view of future classification. Of course, this supposes that the texts have already been read (and understood). Observe that translating a text into a natural language to another one (as, for example, this paper from French to English) requires such a reading and (subtle) understanding of the text.[15]

   One can also choose keywords for a text using totally different criteria. For instance, rather than reading the text itself, one can read and understand an outline or the table of contents and this is also a bottom-up mode. One can also look at an index (if it exists some): a *limit case* which follows a top-down mode. Indeed, no understanding of the words is required to select keywords (though, of course, it does not harm to understand them), one only consider which words are mentioned in the index and their relative importance (which a good index makes clear). Without index, one can also count occurrences of words in a text:

[13]Depending on how much *abstraction* is wanted (or how much *refinement* is wanted), a *text* will be represented by a *binary word* (the blank spaces separating words being also encoded as special characters) or by a *sequence of binary strings* (each word in the text being represented by a string in the sequence). In this paper, we mostly consider encodings of texts with binary words (in particular, for the examples) and not sequences of binary words, and we consider sets of such texts.

[14]It is one way of seeing things! The one reflected by the Anglo-Saxon terminology "top-down". What is essential is that texts are apprehended from the *outside*, in opposition to apprehension from the *inside*.

[15]With a purely syntactic automatic translator, such as the one in Google, one can get results like the following one: "Alonzo Church" translated as "Église d'Alonzo" (i.e. church in Alonzo)!

this is precisely what Google does in its searches. In practice, both bottom-up and top-down modes are often used together (*mix* mode).

Whatever method is chosen, the choice of keywords generally assumes (though it is not always the case) a preliminary knowledge or some general idea of the wanted classification for which the keywords are chosen. This knowledge is a very abstract form of *semantics*,[16] which can evolve through time as new texts are being read. Generally, the person who writes the text is not the one who has this knowledge, this is rather the person who "manages" the classification.

2. Whatever approach was used, once the keywords have been chosen and stored in some way, they give a kind of classification for this text considered among the other ones which have been treated in a similar way. Using given keywords, one can look for all texts which have been assigned such keywords. Clearly, a notion of *query* emerges from the so used keywords. In an extended concept of keyword – and this is exactly how Google works – one can look for all texts containing these keywords (i.e. with these keywords in their contents), with no need o define any keyword for texts.

   Observe that searching using keywords – and this is a fundamental point – is *an apprehension of texts, i.e. an approach to their classification which is top-down.* A set of keywords (a "conjunction" of keywords) plays the role of an *oracle* to grasp texts from the outside, without reading nor understanding them. Using such a set of keywords, one can select some texts among a family of texts which can be really big, even gigantic in the case of the Web. The selected texts can then possibly be read and be understood (at least better understood). One can also group them with other texts having some common keywords and thus get a classification of texts.

3. With the Google approach to classification, things are similar: the choice of keywords for queries to Google (which are in fact conjunctions of keywords) can be done in two ways.

   - In a bottom-up mode this choice comes from the reading and understanding of the content of the Web.

[16]Let us mention that a new concept emerged: that of *thesaurus* which is somehow an abstract semantics related to classification. A thesaurus is a particular type of *documentary language* (yet a new concept) which, for a given domain, lists along a graph words and their different relations: *synonymy*, *metaphor*, *hierarchy*, *analogy*, *comparison*, etc. Thus, a thesaurus is a kind of *normalized and classified vocabulary* for a particular domain. Otherwise said, the words a thesaurus contains, constitute an *hierarchical dictionary* of keywords for the considered domain. One can possibly add definitions of words or consider the classification of words (according to the future usage of the thesaurus) to be sufficient. It is a remarkable tool. First used for disciplines around documentation and for large databanks, it is now used almost everywhere. To build a thesaurus, one follows a bottom-up or a top-down mode or mixes both modes, exactly like in the case of keywords. More details on the notion of thesaurus in the section devoted to databases (cf. Sect. 4.4.2).

- In a top-down mode this choice is based on criteria totally exterior to the content of the Web though it is hard not to be somewhat influenced by previous readings from the Web...

In general, both bottom-up and top-down modes are mixed for the choice of keywords.

Whatever approach was used, once the keywords have been chosen, one has at disposal a kind of oracle to grasp the Web. Otherwise said, *the Google query written with these keywords will select texts from the Web – and also hypertexts or multimedia data: pages from the Web – with a top-down operational mode.* Such selected texts can then be read, classified, etc.

One can also surf the Web along a bottom-up mode, that is give up query and go from one page to another one via the *links hypertext*. Indeed, those links are the main originality of the Web. From a theoretical point of view, they are very interesting since they convey a form of semantics. Thus, the *notion of keywords (and more generally of words) appears to be a limit concept between syntax and semantics.* In general *surfing is done via both approaches : bottom-up with hypertext links and top-down with queries.* As before, the choice of the keywords submitted to Google in view of a classification is also a form of semantics. Observe that in a top-down approach for the choice of keywords, one can however choose them *randomly*, then use some counting (with statistical tools) to get classifications of the selected texts. Such random choices are particularly interesting when there is a huge quantity of texts, which is the case with the Web. However, it is doubtful that such an approach to classification – if fundamentally random – can give significant results. Nevertheless, it can be coupled with a more "deterministic" approach.

If we go back to the title of this section: *Bottom-Up versus Top-Down modes.* It is reasonable to question: why are there two possible modes in the definition of mathematical and computer science objects and in the runs of computer programs? De Facto, these two modes do exist and they are the fundamental modes which have emerged from the works of the diverse researchers in computability theory in the twentieth century. We have seen that these two modes could also be considered in the approach to classification of information and we gave an example with keywords. We have also seen how the use of Google to search the Web was relevant to these approaches. This shows that these bottom-up and top-down modes are not particular to classification: they concern, in fact, any information processing hence any, more or less abstract theory of information. This also concerns all disciplines which deal in some way with the notion of *representation* or *definition*, or *description*, etc. This includes logic, Kolmogorov complexity and computer science, semiotic and also all sciences of cognition: as we detail in Ferbus-Zanda (In preparation-d), *information processing by the human brain could fundamentally be structured around these two operational modes.* In any case, this is quite an interesting approach to cognition which is much enlightened by the evolution of mathematical logic and computer science.

In this paper, we shall look at these bottom-up and top-down modes in two types of situations (concerning classification) which generalize what we said about keywords. Namely,

- The logical formalization of information systems via databases (Sect. 4.4.2).
- The set theoretical approach to the notion of grouping, based on the Zermelo-Fraenkel (ZF) axiomatic set theory. We shall particularly look at the comprehension schema in ZF (Sect. 4.5).

These reflections will help to understand the role played by the Kolmogorov complexity in information classification and more precisely in the notion of *grouping* of informations. We will have to reconsider the notions of intensionality, abstraction semantics and representation in this context (cf. Sect. 4.6). Also notice that the existence of two such modes for the definitions of mathematical and computer science objects, functions and programs and for the execution of these programs, is quite interesting. The fact that we find these two modes for the various forms of the information processing and different disciplines, indicates that this observation is a fascinating scientific project. Clearly these two modes, so *complementary*, form a *duality relation*, a kind of correspondence between two distinct ways of processing which are somewhat distinct and also similar.[17] More precisely, we have seen that the bottom-up approach (on which are based the iterative definitions), results from the notion of set theoretical union whereas the top-down approach (on which are based the inductive definitions), results from the notion of set theoretical intersection. It is therefore quite natural to revisit these approaches in the framework of Boolean algebras, a theory where the notion of duality is typical, so we do in Ferbus-Zanda (In preparation-c). Other fundamental dualities for logic and computer science are also developed in those papers. Especially, duality *syntax versus semantics* and also duality *functional versus relational*[18] which concerns, among others, the relation between algorithms and (functional) programming on the one hand and discrete information system and their formalizations[19] on the other hand. Recall that the essential part of discrete information system is the organization (the structuring) of information,

[17]The abstract notion of *isomorphism* in mathematics is a form of duality. Some dualities are not reduced to isomorphisms. Typically, Boolean algebras with the complement operation (in addition to additive and multiplicative operations) contain an internal duality and are the basis of deep dualities such as Stone duality which links the Boolean algebras family and some topological spaces. The complement operation confronts us to many problems and deep results...

[18]Since Gottlob Frege invention, at the end of the nineteenth century, of the mathematical logic and the formalization of the mathematical language that results from it, mathematicians have de facto to deal with two distinct categories of mathematical symbols: the *function symbols* and *relational symbols* (or predicate symbols) in complement of symbols representing objects. To each of these two large classes of symbols respectively correspond algorithms and information systems.

[19]The information systems in which we highlight a type of programming that we named *relational programming* in a research report: Ferbus-Zanda (1986). We present in this paper the link between functional programming and relational programming.

whatever is their nature (admitting a discrete representation), with the objective of an easily extracting particular informations. Clearly, information system are linked to classification. Thus, we believe it is interesting to present them in this paper. We shall articulate this presentation around the bottom-up versus top-down duality, which is in this way illustrated.

4.4.2 Information System and Database: A Formal Approach

First let us point out that: *databases (*DB*) are to information systems what are computer programs to the intuitive notion of algorithm: a formal, mathematical presentation and the ability of an also formal processing.* Indeed, algorithms and information systems are generally expressed in a natural language (in a more or less clear way) and assume implicit content (which can be important) and also unspoken comment (which may be quite a problem). Recall that algorithms and information systems have existed since Ancient Times.[20] In both cases, this formal expression is essentially done in the framework of mathematical logic. Observe that programming and algorithms are particularly related to lambda calculus whereas databases and consequently information system are particularly related to set theory.

As concerns programs and algorithms, let us mention the remarkable work of Yuri Gurevich (Dershowitz and Gurevich 2008). He introduced a notion of Abstract State Machines (*ASM*), which is based on model theory (in logic) and is a mathematical foundation of the notion of algorithm which is as much as possible refined. Not only does *he captures the notion of algorithm, but he also formalizes their operational mode.* More precisely, Gurevich deals with *operational semantics*, i.e. *the way algorithms and programs are executed*, (the outcome is the programming of an *interpreter and/or compiler* and of an executor of programs). This highly constructive operational point of view completes what is called *denotational semantics* and which deals with what *algorithms and programs compute.*[21] This is, in fact, the way Gurevich states his thesis: ASMs capture the *step by step* of the execution of sequential algorithms. For Gurevich, any given algorithm (in particular, any computer program) "is" a particular ASM which is going to mimic his functioning. This allows to consider an algorithm as a formal object (namely, an ASM). Gurevich's thesis extends Church-Turing's thesis (at least for sequential algorithms): indeed, Gurevich thesis proves it. More precisely, Church-Turing thesis is about denotational semantics (the diverse computation models which have been

[20]Some exhaustive descriptions of algorithms about trading and taxes date from Babylonia (2000 BC to 200 AC). Information systems really emerged with mecanography (end of nineteenth century) and the development of computer science. However, there are far earlier examples of what we could now call information systems since they show a *neat organization and presentation of data on a particular subject*: for instance, the Roman census.

[21]Observe that these semantics correspond respectively to Arend Heyting's semantics and Alfred Tarski's semantics.

imagined are pairwise equivalent: we say they are Turing-complete). Gurevich extends this thesis to operational semantics: ASM are a computation model which is *algorithmically complete* (cf. also Sect. 4.7 and Ferbus-Zanda and Grigorieff 2010). What is really remarkable with ASMs is how their formalization is simple and natural, which, in general, is not the case with the other approaches to operational semantics of computer programs. We come back to ASMs (and their relation with Kolmogorov complexity and classification) in the conclusion.

As concerns, information system (which is an intuitive notion) and their modeling via database (which is a formal approach), we shall see that, historically and conceptually, things were not as simple as they were with programming and the formulation of theoretical models for computability – which, indeed, occurred at a time when there was no computers. In the case of information systems, it was all the opposite. Recall that the first formalization of the representation and treatment of data, (that is what is now called an information system) is Codd's relational model for databases (Codd 1970). What was quite original with Codd's approach is the idea that there were mathematics which should "manage" information in computers. Though this may seem quite obvious now, up to the time Codd created his theoretical model (a time where programs were written on punched cards), that was not the case: computer files were stored in a great mess.[22] One of the most fundamental and unprecedented feature of Codd's relational model is the formalization of the notion of *query*. He founded this notion on a new calculus: *relational algebra* which is a kind of *combinatory logic with operators acting on tables*[23] joined together: classical set theoretic operations (*union*, *intersection*, *cartesian product* and *projections*) and also new operations: *selection* and *join*. It turns out that the join operator is really a fundamental one in logic. Codd also develops a *normalization theory* to handle the very difficult problem of removing *redundancies* in information systems.

Surprising as it is, though Codd worked in an IBM research center, he had to fight very hard[24] to impose his views. The first implementation of his model was not done by IBM but by *Oracle*, at that time a very small company,[25] which saw its exceptional interest and implemented it in 1980. It is only a few years later that IBM

---

[22]*Multics* was the first important operating system to store files as nodes in a tree (in fact a graph). Created in 1965, it has been progressively replaced since 1980 by *Unix*. Derived from Multics, it includes a new feature: multiple users management. Now, all operating systems are based on Unix. Multics was a turning point in the problem of data storage: until now, one speaks of *hierarchical model* and *net model*. But, in fact, these "models" have been recognized as models only after Codd introduced the relational model! Finally, observe that the graph structure of the Web also comes from the organization of files with Multics.

[23]This combinatory logic has much to do with the programming language *Cobol* created in 1959.

[24]The dedication in his last book (Codd 1990) is as follows: To fellow pilots and aircrew in the Royal Air Force during War II and the dons at Oxford. These people were the source of my determination to fight for what I believe was right during the 10 or more years in which government, industry, and commerce were strongly opposed to the relational approach to database management.

[25]Oracle is now a company worthing billions dollars.

also implemented Codd's model. Now, all existing DBMS (database management systems) are based on Codd's relational model. Let us mention that databases are still largely underrated though it could be so profitable in many disciplines. But this is clearly not to last very long due to the dissemination of digital information (with an incredible speed, no one could have expected a few years ago).

There is another theoretical model for databases: the *Entity/Relationship model* due to Peter Pin-Shan S. Chen (1976). This is a formal approach to databases which essentially relies on Codd's relational model but is more abstract. In this model, a database is represented as a graphic which looks like *flow charts* used in the 1960s–1970s to modelize computer programs. It is the source of the language *UML*,[26] which is a system of graphic notations for modeling. In our opinion, Chen's theoretical model is very deep and it should still be the source of many important works. Databases rested on the Entity/Relationship model deserve to be called *conceptual databases*. They constitute an abstract logical extension of relational databases which should have a fundamental role in the future as concerns information processing, classification and any algorithmic information theory. *Object-Oriented Programming Concepts* are also inescapable in information processing and in the approaches to classification. Let us mention the *inheritance* concept in the difficult problem of concurrent access to data, i.e. when the same data is used by several actors: attributes, processes, systems, users. Another important concept from Object-Oriented Programming is that of *event-driven programming*: a particular value in the execution of a program or particular data in a database trigger the execution of some (other) program. Lastly, let us mention another theoretical model for databases: the *deductive model* (also called *deductive database*). This is also a fundamental model. It mixes Codd's relational model and the predicate calculus, bringing intensionality (i.e. abstraction) to Codd's model through the *in extenso* adjunction of first-order variables. The query language for deductive databases is *Datalog*. It is a pity that the existing implementations of Datalog, which work quite well, are only used in some research labs. Currently, there is no "real" deductive DBMS (Database Management System) with the same facilities offered by relational DBMSs. This is quite surprising as information system, with the Web, have taken such a huge impact.

One can also question why diverse theoretical models, as fundamental as they are, can coexist with no serious attempt to mix them. Maybe, this is because database is a very recent discipline, quite probably, this will happen in the near future. We are working towards this goal with the notion of conceptual databases,[27] using logic as a foundational theoretical basis. Consider the general problem of classification of information. Database, with the diverse theoretical models described above, constitute a formal approach to that question. Especially with the notion of query

---

[26]UML (*Unified Modelling Language*) is a formal language, which is used as a method for modeling in many topics, in particular, in computer science with databases and *Object-Oriented Conception* (*OOC*) – in fact, this is the source of UML.

[27] Ferbus-Zanda (In preparation-a).

which becomes a mathematical notion (which, moreover, is implemented) far more sophisticated than keywords. In fact, queries generalize keywords.[28] Whichever theoretical model of database is used, a fundamental primitive notion is that of *attribute* (which can be seen as formal keywords) and different kinds of set groupings of attributes so as to make up the *relational schema* of a database. This constitutes the wanted *formal classification* of the initially unorganized data. The relational schema of a database is the structural part of a database: its *morphology*. So, in the relational model, a database is structured in *tables*. The names of the columns of a table are some attributes for the database. A line in a table describes an entity (from real world): this entity is reduced to the values (for that line) of each of the attributes of the table (i.e. the names of columns). There are relations between the tables of a database, kind of *pointers*, which follow some "diagram" relying on the chosen relational schema of the database.

The content of the tables constitutes the *semantics* (otherwise said, the current *content*) of the database at some particular time. Each table is structured in columns and can also be seen as a set of lines (the so-called "tuples"). The number of columns is fixed but the set of lines varies along time. Each line is a set of values: one value per attribute (recall columns and attributes are the same thing).[29] This notion of line corresponds exactly to that of card in physical files, (for instance, those used to manage libraries in pre-computer days) or to the content of a punched card (mechanography).

For instance, suppose we have a table about authors of books in a library which has the following attributes: `AuthorSurname`, `AuthorName`, `AuthorCountry`, `AuthorTimes`. The `AuthorSurname` column will contain names (such as `Duras`, `Sarraute`, `Yourcenar`, `Nothomb`, `Japp`, etc.)

A typical line could be { `AuthorSurname.Duras` , `AuthorName.` **Mar-**`guerite` , `AuthorCountry.France` , `AuthorTimes.XX th century` } or also the 4-tuple (`Duras` , `Marguerite` , `France` , `XX th century`) since the ordering of values in this tuple makes it possible not to "explicit" the associated attributes.

Queries allow to access these contents. Note that the contents of the tables evolves through time due to *updates* of information: *adjunction, removal, modification*. A database looks like the set of sheets of a spreadsheet (*Excel*) augmented with links between them that are managed through queries (which spreadsheets cannot do, or in a very rudimentary and complex way).

---

[28]One should rather say that keywords – used with web browsers – constitute very elementary database queries (of course, database queries are much older than the Web which emerged only in the 1990s).

[29]Lines are usually presented as tuples but, conceptually, this is not correct: in Codd's relational model there is no order between the lines nor between the columns. Codd insisted on that point. In fact, conceptually and in practice, this is quite important: queries should be expressed as conditions (i.e. formulas) in the relational algebra, using names of attributes and of tables. For example, it means that queries cannot ask for the first or twentieth line (or column).

Thesauruses (cf. Note 16) are, in fact, databases. The relational schema of such a database is the structure of the considered thesaurus, otherwise said, the layout, the architecture of the thesaurus. The diagram of this database (which is a graphic representation of its relational schema) formally expresses this architecture. It is clear that there can be several tables in this database. For instance, in a thesaurus dedicated to the epistemology of mathematics, there could be specific tables for mathematical logic, probabilities, algebra, topology, geometry, functional analysis, differential calculus, integration, etc. and other tables dedicated to mathematicians (mentioning the concepts they introduced), to philosophers, to historians of mathematics, etc. Of course, the choice of such tables is completely subjective. One could structure the database very differently, considering *synonymy, quasi-synonymy, connectivity, analogy, comparison, duality, contrast, etc.* among the diverse words of the thesaurus. The internal organization of a given table (the choice of the columns i.e. of attributes) depends on what one intends to do with the thesaurus and on the choices already made for the diverse tables. The contents of the tables are then constituted by all the words put in the thesaurus.

Without definitions, the thesaurus is a kind of hierarchical dictionary of synonyms, associations, etc., i.e., a structure on keywords. To augment it with definitions, we insert them as contents of the tables in specific columns. In any case, let us stress that the relational schema of the associated databases essentially relies on the "association" part of the thesaurus (indeed, its graph) and not on its "definition" part. Also, observe that it is the power of computers and databases which makes it possible to build and use such complete thesauruses. It would unrealistic to try a readable paper version of a dictionary which would be at the same time a usual dictionary and a synonym dictionary and would also give definitions,[30] but any good computer graphical user interface makes it possible. Note that what is not explicitly represented as a table can be recovered via some query. For instance, if we decided a structure by discipline, one can obtain all synonyms of a given word, whatever be the table of the thesaurus database in which they have been inserted (according to their associated discipline). This shows that any particular choice of a structure for the database leads to no disadvantage as concerns the usage of the database: whatever grouping of information is wanted, it can be obtained via some appropriate query. This is "hidden" to most users which have no idea of the internal organization of the database. In general, one chooses a structure which makes easier the elaboration of the schema of the database, or an *optimized* structure to get efficient executions of

[30] In fact, any such "complete" dictionary is necessarily circular: a word a is defined using the word b which is itself defined with other words themselves defined in the dictionary. It requires some knowledge *external* to the dictionary to really grasp the "meaning" of words. Note that this incompleteness is more or less "hidden". On the other hand, in a synonym dictionary, the structure essentially relies on circular definitions. This is less apparent with paper dictionaries: for a given word, there will be only references to its synonyms. However, with digital dictionaries, this *circularity* is really striking: links "carry" the reader to the diverse synonyms and can be implemented with pointers.

queries (recall there are database tables containing millions of lines). Of course, the synonymy in question is relative to the closed world of the database formalizing the thesaurus.

The result of a query in relational databases is a *view* which is structured as a table. The only difference between a table and a view is that views are stored in the RAM (random access memory) of the computer (which is a volatile memory: it disappears when the computer is turned off) whereas "real" tables of the database represent *persistent* data which are stored on non-volatile memory: hard disks, magnetic tapes, etc. Of course, one can nevertheless save a view.

Observe a very interesting phenomenon with this example: the emergence of the notion of database. Indeed, following the same approach, one can build a database dedicated to epistemology of physics, of chemistry, of biology, etc. and group these databases in a unique database in order to get a thesaurus dedicated to epistemology. One can also group epistemology with other disciplines. Clearly, one has to fix the *wanted level of abstraction/refinement* to build the thesaurus (or, more generally, a database) and what is the limit to the considered subject. This is one of the most difficult problems in modeling. Any scientific activity goes along a particular answer to that problem. This example leads to the following observation: in this paper, the notion of "object" has not been much considered. It is clear that the *hierarchical* character on a thesaurus relies on *inheritance* (a concept from OOC, cf. above). It seems therefore necessary to add to Codd's relational model some concepts of the object oriented approach,[31] which is what we try to do with conceptual databases.[32]

If we consider the general case, we observe that *the notion of query in databases is essentially dependent on the structure of the database associated to the relational schema.* Database queries are similar to Google queries with one big difference: queries in relational database are written in a programming language which is far more sophisticated than conjunctions of keywords allowed in Google queries. In all implementations[33] of Codd's relational model for databases, queries are written in the programming language *SQL* (*Structured Query Language*).

As with keywords, the choice of attributes and that of groups of attributes in a database is completely subjective: this is *semantics* and this semantics is formalized by the relational schema. Once such choices are done and the relational schema is fixed, the form of possible queries is somewhat constrained but, nevertheless, it is possible to ask whatever is wanted. This was argued above with the example of the

---

[31]Codd was strongly opposed to any addition from the object approach to the relational model. Indeed, the so-called "First Normal Form" (due to Codd) formally forbids the possibility of an attribute structured as a list, a tree or a graph (which is exactly what OOC would do). When he elaborated his model, this was a reasonable choice: the object approach is quite destructuralization while Codd's approach was a structuralization one. Let us mention that Codd also opposed Chen's Entity/Relationship model (nobody's perfect)!

[32]Ibid. Note 27.

[33]An implementation of Codd's relational model for databases is a DBMS (*DataBase Management System*). Any DBMS includes an interpreter of the language SQL (such an interpreter is, in fact, an implementation of Codd's *relational algebra*, the fundamental calculus in this theoretical model).

thesaurus. As for the Web, such a relational schema is absolutely impossible because the Web is so fundamentally dynamic.

Observe that, at any step, we have with databases a precise idea of the structure we are working on (it is a mathematical object) and extracting information out of such a structure is done in a rigorous way, using the formal notion of query.

Let us then notice that the result of a query is *exhaustive* relative to the database we consider: we get exactly all objects in the base that satisfy the query, *no more no less*. Also notice that the information content of a (correctly formalized) database is precisely known at any time and the modifications brought to the base (adding, removing or changing data) is precisely controlled. Of course, this is not the case when extracting information from the Web with a search engine and this is not the case either for large data banks (in biology, medicine, cartography, etc.) which have no solid mathematical foundation as have relational databases neither in the structuralization of data nor for the queries. Databanks are indeed databases which are somewhat not well formalized (or somewhat ill). In other words databanks can be really databases whereas this is intrinsically impossible for the Web.

4.4.3 Database and Bottom-Up Versus Top-Down Duality

Let us now look at the elaboration and use of databases in the perspective of bottom-up and top-down approaches. It turns out that this is much the same as with keywords and Google queries.

1. The choice of the relational schema is done using a bottom-up or top-down operational mode. In general, both modes are used jointly (in fact, alternatively). In the bottom-up mode, one uses the expected future content of the database to build its relational schema (which will structure this content). In the top-down mode one builds the relational schema on considerations which are external to the future content. At first glance, using the bottom-up operational mode may seem paradoxical: to use the content in order to structure it. But this is not the case. In practice, to build a relational schema for a given database, one starts from some sketchy idea of the schema, represents it as some graphic (top-down approach), then implements it (this is programming work). A kind of *prototype* is thus obtained. This being done, one fills the tables of the database with a few lines (a "set of data") to test the pertinence of the relational schema, which may lead to adjust it (bottom-up approach). And this may be repeated... Recall that the content of a database is precisely what gives the semantics of the database whereas the construction of the relational schema is morphology (syntax). With such a mix approach, one can build the *morphological (syntactic) part of the database via some access to a part of the semantics of the database. And vice-versa.* Thus, this approach, so seemingly paradoxical, is not so. In fact, there are two true difficulties. First, to delimit the scope of the information system which is to be modeled, and this is done using the given *specifications*. Second, to choose the right level of abstraction of each component (attributes, tables, etc.).

2. The choice and programming of queries comes next. And the approach is bottom-up, top-down and mix: this is similar to what we said about the elaboration of the relational schema. However, for quite complex databases, one may have to build the schema and the queries more or less simultaneously: we saw this with the thesaurus.
3. Once the relational schema of a database seems adequate and the main queries have been written down and programmed (some of them testing the coherence of the base), one can really fill the database and complete its content. Queries can be added as wanted. But any modification to the relational schema, even a seemingly minor one, can cause a great damage when the size of the database is somewhat huge. For instance, breaking an attribute `Artiste` into two attributes `Composer` and `Interpreter` in a music database.
4. *The content of the database can then be grasped through a completely top-down mode using queries.* This is why relational databases are such a breaktrough. Huge quantities of data can be accessed from the outside in a completely rigorous mathematical way. The notion of query can then be seen as a kind of *oracle*. Of course, one can also follow a bottom-up approach: browse the content of the database to find some wanted information. Before Codd's relational model, this was, indeed, the sole possible approach (excepted mechanography) with the old physical "files" such as index cards in large libraries: alphabetical (syntactic) sorts caused no problem but sorting such files according to themes (semantic) was a real headache!

4.4.4 Classification and Bottom-Up Versus Top-Down Duality

Let us summarize. Approaches to classification via keywords or via Google queries (such as Google classification), databases (whatever theoretical model is used) have the same intrinsic nature. *In the diverse phases of the elaboration, especially with keywords and queries, one can follow a bottom-up operational mode or a top-down one* (and generally, both modes are used alternatively in a *mix* mode). *Queries obtained in that way then allow to grasp sets of texts in a top-down mode (that is with no understanding of the meaning of the texts) and classify them. The approach to classification using compression is entirely relevant to the top-down mode.* Observe that, for the classification using compression, the framework is then *purely syntactical*, there is no use of any keyword or query which would convey some semantics (for instance, that given by the chosen identifiers). *Thus, one gets information relative to texts without turning to their semantics: simply compress and compute.*

At first glance, this approach may seem somewhat "miraculous": one is able to classify information contained in texts without getting into their contents and with no need to understand them. On the contrary, in the previous approaches, one is lead to use a bottom-up mode (though this is not absolutely needed) to build interesting queries (and the relational schema in a database). Let us recall what we evoked supra: text compression is a highly theoretical science and a simple, current-use algorithm such as "gzip" is the result of years of research. Of course, in classification by compression, texts are not chosen randomly! However, for the next future, one sees no limit to the usage of the above method to all information which is on the Web. Considering the general problem of classifying information, observe that *statistics* constitute a particular case. Usually, the statistical approach is top-down, computing *correlation factors* to group objects and/or informations and get a structure on them. Indeed, Google and compression algorithm heavily use statistics. Nevertheless, one can also follow a bottom-up mode with statistics or even mix these two approaches. This will be seen below where we propose a probabilistic version of the comprehension schema (cf. Sect. 4.5.2).

4.5 Set Theory Interpretation of Bottom-Up Versus Top-Down Duality

Let us now look the different approaches to classification in the perspective of the comprehension schema in Zermelo-Fraenkel set theory **ZF**. A theory which can be viewed as one of the first formal mathematical attempts to approach the notion of classification, sets being the most rudimentary way to group elements. As a matter of fact, Codd's relational model for databases relies on (naive) set theory, which is not so surprising in the search of a formal structuralization mode. Thus, the *bottom-up versus top-down duality* that we point in classification (cf. Sect. 4.4), can be illustrated by the way the set theoretical comprehension schema "works". We also discuss a probabilistic version of the comprehension schema which among others illustrates the *exact versus approximate* duality.

4.5.1 The Set Theoretical Comprehension Schema

This is an approach from "pure" mathematics.

It is a global approach, intrinsically deterministic, going along a fundamental dichotomy:

True/False,
Provable/Inconsistent.

A quest for absoluteness based on *certainty*. This is reflected in the classical comprehension schema

$$\forall x\ \exists y \quad y = \{z \in x \ ;\ \mathcal{P}(z)\}^{34}$$

---

[34] More formally: $\forall x\ \exists y\ \ \forall z\ \ (z \in y \longleftrightarrow (z \in x \wedge \mathcal{P}(z)))$.

where $\mathcal{P}$ is a *known property fixed in advance*. Thus, the set clustering is done from a well known property which is defined within this dichotomy. *To do such a grouping and build such a set, we again find ourselves in top-down operational mode: this set is being constructed from the property $\mathcal{P}$.*

More precisely, with a constructivist approach:

- We start with a set x.
- We choose a property $\mathcal{P}$ relative to elements of the set x. This can be done in both bottom-up and top-down modes exactly as in the choice of keywords for a query or as in the elaboration of a query in a relational database (cf. Sect. 4.4.3). Note that the idea of the grouping, i.e. the *choice of this grouping (formalized by the property $\mathcal{P}$)* is completely subjective: this is *semantics*. Nevertheless, we can also get such a property $\mathcal{P}$ in a syntactic way: through a computation (cf. Sect. 4.6.1).
- Having this property $\mathcal{P}$, we then pick the elements of the set x which satisfy $\mathcal{P}$.

The comprehension schema[35] allows us to consider such a set construction (in the ZF axiomatic set theory). If we do not relativize this construction to some fixed set x (or, equivalently, if we consider a set containing all sets) then we face Russel's paradox.[36] Observe that the solution to this paradox really makes sense: in this approach, one should start from something, and it will be from an existing set to work with such a property! Indeed, the elaboration of the property $\mathcal{P}$ is made in a mix mode (as with queries in a relational database) then we can start with a certain idea for the property $\mathcal{P}$ (related to what is the set x) then "pick" some elements in the set x to get a better idea of $\mathcal{P}$, and then pick again some elements in x and adjust $\mathcal{P}$, and so on.

Once this property has been "set up" (maybe getting it *in extenso*), one is now able to group all elements of x which satisfy $\mathcal{P}$. Of course, in the mathematical literature, no one present such successive approximations to get a property: the obtained property is given directly! Nevertheless, this is how things are being done in general. Computer scientists are used to such practice: a modular approach is used to perfect a database or a program. Of course, so do the mathematicians quite often. It is important to note that the grouping, that is, the definition of the set y or its constitution (though some would rather consider an explicit construction) can be done in a *top-down operational mode which is an intensional mode. Intensionality, (one can also say abstraction) is expressed by that property $\mathcal{P}$ which plays the role of an oracle for the set y, and is the exact opposite of an extensional description (which gives the element, one by one) which is necessarily done in a bottom-up mode.* Knowing in advance the property $\mathcal{P}$ is a very particular case which does not happen

---

[35]One can also constraint in different ways this property $\mathcal{P}$. In particular, to avoid circularities such as the one met when $\mathcal{P}$ contains some universal quantification on sets, hence quantifies on the set it is supposed to define (this was called *impredicativity* by Henri Poincaré).

[36]Russel's paradox insures that the following extension of the comprehension schema is contradictory: $\exists y \quad y = \{ z \ ; \ \mathcal{P}(z) \}$, i.e. $\exists y \ \forall z \ (z \in y \longleftrightarrow \mathcal{P}(z))$. Indeed, consider the property $\mathcal{P}$ such that $\mathcal{P}(u)$ if and only if $u \notin u$, then we get $y \in y$ if and only if $y \notin y$.

in most "real" situations. Below, we develop this aspect by proposing a "probalistic" comprehension schema. Then we show in Sect. 4.6, how this probabilistic schema can be generalized using Kolmogorov's complexity. This brings us to the relation between the algorithmic information theory and classification which are the heart of this work.

4.5.2 The Probabilistic Comprehension Schema

In the probabilistic approach, much more pragmatic than the logical one, *uncertainty* is taken into consideration, it is bounded and treated mathematically.[37] This can be related to a probabilistic version of the comprehension schema where the truth of $\mathcal{P}(z)$ for instances of z is replaced by *some limitation of the degree of uncertainty of the truth* of $\mathcal{P}(z)$. Formally, together with z, we have to consider a new parameter in $\mathcal{P}$, namely the event ω of some probability space Ω and we have to fix some confidence interval I of $[0, 1]$ (representing some prediction interval). Denoting by μ the probability law on Ω, the probabilistic comprehension axiom for property $\mathcal{P}$ now states

$$\forall x\, \exists y \quad y = \{z \in x \;;\; \mu(\, \{\omega \in \Omega \;;\; \mathcal{P}(z, \omega)\}\,) \in I\}$$

As was the case with the set theoretical comprehension schema, *one gets in a top-down operational mode to do such a grouping and build such a set* from property $\mathcal{P}$ and interval I. This is so even if we allow some degree of uncertainty for the truth or provability of property $\mathcal{P}(z)$ (which is then replaced by $\mu(\{\omega \in \Omega \;;\; \mathcal{P}(z, \omega)\}) \in I$) for particular instances of z. Once again, this is a precise particular case: though its truth has some uncertainty, this property *is well defined and fixed in advance, together with the confidence interval I*. However, such a schema is closer to many situations met in the real world. As in the previous case, such a property $\mathcal{P}$ (and the confidence interval I) allow to *define the set y in a top-down operational mode, that is to get an intensional, abstract description of the set y*. It is natural to consider property $\mathcal{P}$ as a kind of *oracle* with non totally accurate answers (the interval I limiting the inaccuracy). Observe that, as above, the *choice* of $\mathcal{P}$ and I is relevant to *semantics*. Remark that there are other ways to formulate a probabilistic comprehension schema.

As concerns groupings of information relevant to a purely top-down mode (the grouping itself, the elaboration of a property to do it, the definition of sets of information), we treat it in the next Sect. 4.6 about intensionality and Kolmogorov complexity. Let us simply recall (cf. Sect. 4.4.4) that classification by compression and some methods based on statistical inference allow to have such purely top-down approaches. The particular of Google classification is exactly the same as that

[37] We refer the reader to Feller (1968) and also Kolmogorov (1956, 1983), and Chaitin (1969).

of set theoretical and probabilistic comprehension schemas (for Google, keywords play the role of a property $\mathcal{P}$) and that of classification via databases, up to one significant exception: *with Google, everything is moving: answers as well as the keywords proposed in queries.*

4.6 Information, Intensionality, Abstraction and Kolmogorov Complexity

4.6.1 Classification, Database, Intensionality, Abstraction, Semantics and Algorithmic Information Theory

We stressed in Sect. 4.4 the importance of the Web expansion and the huge interest of classification by compression and Google classification. The Web can be seen as a phenomenal *expert system*: first, it is a huge information system (this is the network aspect, software and hardware, between machines and servers), second, machines are used and programmed by human beings (their brains) with far more intelligence than what is done in the syntactic world of machines which can only compute. Classification by compression (and Google classification) will surely be more and more used with information on the Web. The same is true with statistical inference methods. In some sense, all these approaches are tightly correlated and, as any approach to classification (cf. Sects. 4.4.4 and 4.5), they lead to *top-down approaches to information*. In particular, *they can be used to grasp the information content of a text (and more generally of a set of texts) with no access to it "from the inside", i.e. without reading and understanding the text.* These methods look for analogies with other texts, the meaning of which is known, or they compare their respective information content. Somehow, they are "profilers" which will become incredibly efficient in the near future when applied to information on the Web.[38]

However we have also explained how these methods still lack some formal development, in particular for the notion of query: for any classification of information, the first question is to find back information from this classification. It is a fact that the notion of query to the Web (with Google or any browser) is still not really formalized.

We have seen that Codd's relational database model led to a completely mathematical structure and processing of the information contained in computer files through the relational schema and the possible queries to the database (the scope of such queries being tightly dependent of the relational schema). As said above, before Codd, there was no such information processing with machines. Codd had to fight to impose his mathematical model and, even today, operating systems do not really use databases. A reflexion about possible formalizations of classification

[38]Recall that once an information has been put on the Web, it is almost impossible to remove it...

by compression, Google classification and a notion of query to the Web, is, in our opinion, quite fundamental. Note that with Google (or any other browser) we have no idea how to measure the degree of uncertainty of Google's answers. The percentage of pertinent answers may be anything between 0 and 100 %. Google answers are *unpredictable* and *constantly moving*. Not an easy situation! However, it seems reasonable to ignore at first the moving character of Google (and also its not completely scientific features, cf. Sect. 4.3.2, point 4) when looking for a mathematical modeling of these methods.

Indeed, one starts from a clustering or more generally from a classification, obtained by way of conjunctions of keywords which are proposed into queries for Google or from a clustering or a classification obtained by compression or observed by way of the statistical methods.

In the simple case of a clustering, we infer the existence of a property, of a "law", which is a form of regularity. The emergence of such a law coincides with the existence of a certain degree of intensionality *in the clustering we accomplish. Otherwise said, we make obvious a grouping of objects, the description of which can be compressed by using this property. This is an intensional description (when the compression have been performed). This can be seen as an (extended) top-down version of the set theoretical or probabilistic comprehension schema: the property used in the set groupings is not known and fixed beforehand.*

For more sophisticated classifications, one will have *higher order clusterings*, i.e. clusterings of clusterings, etc. Otherwise said, several properties will be involved (in some cases, even infinitely many properties, in a theoretical point of view). Observe that, using relational databases one can see that, up to now, quite a few levels suffices to modelize a lot of discrete information systems (for the "real world)". One can expect a similar situation for classifications obtained via the top-down approaches as evoked above, at least for those relative to the present real world. Observe that, with a subtle analysis of modelization using relational databases one can see that, up to now, quite a few levels suffices to modelize a lot of discrete information systems (for the "real world)". One can expect a similar situation for classifications obtained via the top-down approaches as evoked above, at least for those relative to the present real world. This points out the remarkable pertinence of Kolmogorov complexity theory which is an avant-garde theory. Especially when being considered with several points of view, namely by studying the randomness of a word or its information content or the possibility to compress this word. Somehow, randomness is the "opposite" of classification, More precisely, there is a duality *randomness versus classification*, coming from the fact that Kolmogorov's theory of algorithmic information allows to look at these two sides of information (this is what Kolmogorov explicitly tells in Kolmogorov 1965).

This duality is a quasi-opposition though randomness is not *chaos* (cf. Part I). This points out deep relations between Komogorov complexity and relational databases (which constitute, up to now, as we saw, the sole implemented – and widely spread – logical approach to information systems). This complexity also appears unavoidable as soon as one is interested in classification problems. This is not surprise since Kolmogorov complexity is primarily a theory about information!

Let us go back to Kolmogorov's approach, one can observe that it is relevant to the *top-down mode*. Indeed, look at the basic definition of Kolmogorov complexity: *The length of the shortest program which outputs a given data* (the output being a binary word which represents a given object).[39]

Larger is the Kolmogorov complexity of an object, larger are all programs to produce it, more random it is, larger is its information content, more incompressible are all programs to produce it, less intensional is any description of it, less intensional is it itself, less abstract is any property that allows us to describe the object (when we consider the property in a syntactical perspective).

In this definition one does not enter into the content of the output or into the details of the object, which is therefore taken as a whole. *One solely handles the object from the outside* via *some program and/or some property which allows to describe it.* This is indeed a top-down approach as are classification using compression, classification using Google and a part of statistical inference methods. And this suggests that these classifications methods are somewhere related and that Kolmogorov complexity could give an unifying mathematical formal framework. In other words, thanks to Kolmogorov theory, we are able to measure the complexity of an object (in the sense of Kolmogorov), i.e. to give a numerical measure of the *degree of intensionality* or even of *degree of abstraction* which is contained in a computable description of that object. It is remarkable that this can be done with no prerequisite "knowledge" of the structure of the object and that this is indeed what allows us to apprehend this structure.

4.6.2 Kolmogorov Complexity and Information Theories, Semiotics

Let us now compare the diverse ways to approach the notion of information followed by Shannon (cf. Part I), Kolmogorov, Codd and other researchers.

- For Shannon (1948), an information is a *message* which is transmitted through some physical device. In particular, an information is a signal and there can be losses during the transmission. This design is that of a *dynamic information approach* and the physical communication medium is of outmost importance.

  So Shannon looks at robustness of information and comes to a quantitative notion of information content in transmitted messages. To measure variation of this quantity, he borrows to thermodynamics the concept of *entropy* and he bases his theory on it. So he clarifies, on mathematical basis, how to deal with noisy communication channels. In Shannon's theory, words represent information (messages). It is based on coding letters or groups of letters in a

[39] $K_\varphi(y) = \min\{|p| : \varphi(p) = y\}$ where $K_\varphi : \mathcal{O} \to \mathbb{N}$ where $\varphi : \{0,1\}^* \to \mathcal{O}$ is a partial function (intuitively φ executes program p as a LISP interpreter does) and $\mathcal{O}$ is a set endowed with a computability structure. We take the convention that $\min \emptyset = +\infty$ (cf. Part I).

word (cf. Partie I), i.e. *it is a purely syntactic analysis of words (and messages they represent) which makes no use of any semantics.*

Thus Shannon elaborates a mathematical theory of the information content of messages transmitted with some loss of signal. Its main (and hugely important) applications are related to telecommunications (no surprise: Shannon worked in Bell Laboratories).

- The origin of Shannon's work is Wiener's cybernetics (cf. Note 7) in the late 1940s. This subject was much discussed in the Macy conferences (New-York, 1942–1953), to which Shannon attended. Before Wiener and these conferences, there was nothing like an information theory.

  Cybernetics is a theory which establishes, among other things, the concept of *auto regulated system*, in terms of: *global behavior*, *exchanges*, *communication* and *interactions*. Fundamentally, this is a top-down approach to information and systems. Wiener talks about *a science of relations and analogies between (living) organisms and machines.*[40] In particular, he studies *random processes* and the *"noise"* occurring during the exchanges in a system. A fundamental notion in his theory is that of *feedback*: An object is controlled by the instantaneous error margin between its assigned objective. This is clearly a prefiguration of Shannon's information theory (Shannon attended Wiener lectures as a student).

  Wiener has an avant-garde vision on *machines*! His works are the origin of many discoveries, in particular, in sociological, psychological and biological aspects of *communication* and *interaction* and, more generally, in all information theories. Besides several research themes generated by Wiener's theory, let us also mention that Wiener's theory has a deep influence on a large part of modern *semiotics*.[41]

- For Kolmogorov (1965) (see also Gregory Chaitin 1966; Ray Solomonoff 1964), the fundamental aspect of information is the *information content* of an object, independently of any consideration on how this information is used (as a message for instance). This is a *static vision of information*. What Kolmogorov is interesting in is to give mathematical foundations for the notion of *randomness* and to explicit the notion of *information content* of a given object which is *intrinsic* to that object. Thus, what Kolmogorov looks for is a mathematical theory of information which would be far more abstract than Shannon's one and would be based on semantics not only on a "physical" object like a word. His solution is to consider computer programs (considered as computable descriptions) – considering things in fact in the context of the calculability theory – which output an object and look at the length of a smallest one. Thus, considering both programs and what the program does, the algorithmic information theory created by Kolmogorov has both syntactic (length of a program) and semantic features (i.e. what the program does).

[40] Wiener's book (1948), raised many controversies (and Wiener exchanged a lot with von Neumann about it).

[41] A subject going back to Charles Sanders Pierce (1839–1914).

With Kolmogorov complexity, one can capture an "objective" mathematical measure of the information content of an object. Moreover, this measure is really inherent to the object – in some way it is an *universal* specification of the information content of the object – since it does not depend (up to a constant) on the considered programming language to get programs: this is the content of *Kolmogorov's Invariance Theorem*. In order to aim an "absolute" mathematical notion of randomness, Kolmogorov makes a drastic abstraction from any physical device to carry information. In this way, he elaborates the algorithmic information theory which allows to "compute"[42] Kolmogorov complexity of any object. Introducing a conditional version of Kolmogorov complexity, he refines this notion of intrinsic complexity of an object by *relativizing it to a context* (which can be seen as an *input* or an *oracle, etc.* for the program) carrying some extra information. This exactly matches the problem pointed by Eco about the necessity to distinguish signification and information content.

This is how Kolmogorov founds *algorithmic information theory*, which can be looked at as much as a *mathematical foundation of the notion of randomness* than as a *mathematical foundation of information classification and structuralization.*

- As seen above, for Codd (1970), the fundamental feature of information is its *structuralization* – which is formally described – and the fact that one *can get back information from this structuralization in an exhaustive way.* Codds theory essentially relies on mathematical logic. Thus, Codd bases his work on the static aspect of information Observe that, as Kolmogorov does, Codd also makes abstraction of the physical device carrying the information. This was quite a revolution in information treatment at IBM: previously, any information treatment dealt with the *files* containing the data: information and files were considered as a whole.

  Observe that the modeling of information systems via relational databases also takes into consideration the subtle distinction between semantics and information content: *the pertinence of an information with respect to a given information system* is seriously considered. The same distinction is taken into account in the construction of the relational schema of a database. For instance, in a database to manage a university, a choice is to be made: is the information about the students hobbies to be considered or to be ignored? Of course, this choice is completely subjective, this is semantics. If the attribute `StudentHobby` is retained then it will appear in the relational schema of the database, i.e., in the *syntactic counterpart of what is retained as the "constitutional" semantics of the information system.*

[42]Recall that the very original idea on which Vitanyi based the classification using compression is to compute an *approximate value* of this complexity via usual compression algorithms.

4.6.3 Algorithmic Information Theory, Representation and Abstraction

A priori, Kolmogorov complexity does not apply directly to the objects we consider, but only to binary words associated to a chosen representation of objects. However, for the usual different representations, this has quite a minor incidence (this is the content of the *invariance theorem*). Thus, we (abusively) speak of the Kolmogorov complexity of objects instead of *Kolmogorov complexity of representations of objects*. Nevertheless, if higher order representations are considered, this is no more true. For instance, if we represent integers as cardinals of (finite) recursively enumerable sets. Indeed, Kolmogorov complexity allows to compare higher order representations of integers, leading to a proper *hierarchy* of natural semantics for integers (*Church iterators, cardinals, ordinals,* etc.) as we proved in Ferbus-Zanda and Grigorieff (2006). This hierarchy can be put in parallel with a hierarchy of Kolmogorov complexities obtained by considering infinite computations and/or oracles. We show, among other things, that Kolmogorov complexity is also useful to get a kind of classification of semantics for integers which is rather amazing. We can also see this classification of different representations of integers as a *classification of the degree of intensionality of these representations, i.e. a sort of classification of the less or more abstract nature of different definitions of integers, obtained from the different semantics we consider.* We develop this in Ferbus-Zanda and Grigorieff (2006).[43]

4.7 Conclusion

The previous considerations show, in particular, that not only Kolmogorov complexity allows a mathematical foundation of the notion of randomness, but this theory is also intrinsically related to the fundamentals of information: the notions of *information content* and *compression*, that of *classification* and *structure*, and more generally, *database* and *information system* (as they currently are). This theory is also related to the notions of *intensionality* and *abstraction*, and also to the notions de *representation*, *syntax* and *semantics*. An enormous scope! This double aspect (randomness and classification) – drawn by Kolmogorov since the origin of his theory (Kolmogorov 1965) – is partly stressed by the denomination *algorithmic information theory* commonly used to distinguish Kolmogov complexity theory and Shannon's *information theory*. Many applications can be expected in various unsuspected domains. And this theory seems to us particularly suited to provide a *unifying theoretical framework* for a lot of approaches to information processing.

---

[43]As in the two forthcoming (technical) papers: Ferbus-Zanda, M. & Grigorieff, S. *Kolmogorov complexity and higher order set theoretical representations of integers* and Ferbus-Zanda, M. & Grigorieff, S. *Infinite computations, Kolmogorov complexity and base dependency.*

However, it seems to us to be interesting, to look for an extension of Kolmogorov complexity. As it is now, it is essentially based on the theory of *computable functions* hence on *algorithms*. What we propose is to extend it by considering *sets*, *information systems* and *databases*. This would put forwards a *relational, non deterministic* point of view which would be in contrast with the *functional*, essentially *deterministic* current point of view, first considered by Kolmogorov himself (this goes along with a new look to ASMs in the relational framework). It would then be possible to revisit (and to increase) Kolmogorov complexity and ASMs in terms of the duality *functional versus relational* (see Sects. 4.4.1 and 4.4.2).[44] This means that we look at Kolmogorov complexity with a more refined and *more structured* point of view – in other words with a *qualitative* point of view – than that of Kolmogorov. For him a program and an output are binary words (which can represent sets, graphs, information systems, etc.) and his main purpose is to get a *quantitative definition of the complexity of an object.*

Such a qualitative approach was also followed by Codd himself while he elaborated the relational model for databases. His theory is based on the formal notion of attribute which is to represent qualitative characteristics of objects (which are related via diverse links which are also of qualitative nature) and Codd puts such attributes in a mathematical framework. A database is a formal and mathematical specification as "scientific" as any algorithm which processes data and computes. In particular, one can look at the smallest program which outputs some given object rather than at its sole length or also look at the set of all programs which give the wanted output. Such an approach enlightens new links between algorithmic theory of information and Gurevich's ASMs.[45] It opens promising perspectives. As Gurevich told us,[46] the ideas of Kolmogorov complexity theory are far from having exhausted all possible applications: it is just the beginning... Classification of information by compression and Google classification witness such new possibilities. It is also in such a structural perspective that Bennet developed the logical depth complexity (Bennett 1988) which considers the running time of the program which gives the output. It is also called the *organized complexity*. Keeping the same spirit (with such a level of refinement), comes this question: *Why consider the shortest program? What is so particular with it?* The answer comes from the observation of ASMs and the Curry-Howard correspondence: *The shortest program is the most possible abstract.* Indeed, Curry-Howard correspondence insures a correspondence between logic and λ-calculus – in that sense, this correspondence is an isomorphism – hence by extrapolation logic and computer programming. Curry-Howard correspondence plays a fundamental role in the articulation of proof theory, typed lambda calculus, theory of categories and also with models of computing

[44] We study the duality of functional and relational in Ferbus-Zanda (In preparation-c). The relation between ASMs and Kolmogorov complexity and the reconsideration of these theories with a relational point of view are developed in a forthcoming paper: Ferbus-Zanda (In preparation-b).

[45] This is what we started in Ibid. Note 44.

[46] Personal communication while he was visiting our university in Paris.

(either theoretical or implemented ones like programming languages). It was known by Curry for combinatory logic as early as 1934 and for Hilbert proof systems in 1958. It was extended by William Howard in 1969 who published a corner-stone paper[47] in 1980.[48]

Let us say briefly that in the Curry-Howard correspondence, one consider that:

- Logical formulas correspond to types in typed λ-calculus and to abstract types in computer science.
- Cut elimination in a proof[49] corresponds to normalization by diverse rules in λ-calculus, including β-reduction[50] relating λ-terms and runs of computer programs.

This enhances the abstract character of programs evoked above. Indeed, *the smallest logical proof (considered in a given context) is in fact the one which contains the most numerous* cuts. We saw (cf. Note 49) that in some cases, a cut is a form of abstraction. Notice that a proof, of which we have eliminated cuts (which therefore means in some situations replacing "a general case" by a lot of "particular cases"), has its size bounded in the absolute by a "tower of exponentiations"... *The more cuts a proof contains the more abstract it is. Somehow, we can say that the more abstract is a proof, the more compressed it is.* In the same way, *The more redexes there is in a λ-term,*[51] *The more abstract is a λ-term, the more compressed it is.* And for computer programs, the notion of cut can also be defined for programming languages

[47]Howard (1980).

[48]Joachim Lambeck also published in the 1970s, about this correspondence concerning the combinatorics of the cartesian closed categories and the intuitionist propositional logic. Note that Nicolaas Debruijn (*Authomath system*) and Per Martin-Löf had also a decisive influence upon the original Curry-Howard isomorphism. Martin-Löf saw the typed lambda calculus, which he was developing, as a (real) programming language (cf. Martin-Löf 1979). Similarly, Thierry Coquand elaborated the *theory of Construction* on which is based the *Coq system*, initially developed by Gérard Huet at the INRIA (France) in the 1980s. (See also Note 50.)

[49]The notion of *cut in the Sequent Calculus and the Natural Deduction* is a fundamental notion in proof theory. It was introduced by Gerhard Gentzen in the 1930s – and these two logical calculus too. In some cases one can see a cut as a form of abstraction where a multiplicity of particular cases are replaced by a general case. In the sequent calculus, a cut is defined by means of the *cut rule*, which is a generalization of the *Modus Ponens*. The fundamental result of Gentzen is the *Hauptsatz*, which states that every proof in the sequent calculus can be transformed in a proof of the same conclusion without using this cut rule.

[50]In fact, Church's original λ-calculus can be extended with constants and new reduction rules in order to extend to classical logic with the notion of *continuation*, Thimothy Griffin, 1990 – and possibly classical logic plus axioms such as *the axiom of dependent choice* – the original Curry-Howard correspondence between intuitionist logic and usual typed λ-calculus. This is the core of Jean-Louis Krivine's work who introduced some of those *fundamental constants* which have a deep computer science significance (cf. Krivine 2003).

[51]A redex in a λ-term t is a subterm of t on which a one-step reduction can be readily applied, for instance, with β-reduction, this is a subterm of the form $((\lambda x.u)v)$ and it reduces to $u[v/x]$, which is the term u in which every occurrence of x is replaced by v (some variable capture problems have to be adequately avoided).

with their usual primitive instructions. For instance, a program containing: `for i = 1 to 1000000 do print(i)`, is more abstract than the same program in which this loop is replaced by the sequence of instructions: `do print(1) and do print(2) and ... and print(1000000)`. Thus, the `for` loop allows for cuts. Hence a result similar to those precedents: *The more cuts a program contains, the more compressed it is.* Observe that the more a program is compressed via cuts, the more *declarative* is this program. Which means that its text contains less *control* instructions, i.e. less instructions about the technical way some parts of the program are to be executed. A fully compressed program is totally declarative.

But what about ASMs in this context?

As we said, ASMs allow to represent- in a very simple way – the step by step of the execution of any sequential algorithm using models in first-order logic and some simple primitive instructions. As can be expected, it is interesting to look for a notion of cut in the ASM framework. In the same vein, deep relations exist between ASMs, λ-calculus and Curry-Howard correspondence. Cf. our paper to appear in honor of Yuri Gurevich (Ferbus-Zanda and Grigorieff 2010), in which we represent ASMs in λ-calculus, showing that λ-calculus is *algorithmically complete* as are ASMs.

Going back to the information context, we can say: *The shortest program producing a given output is the most abstract one, hence (viewed in λ-calculus) it is the λ-term containing most redexes, hence also (viewed in proof theory) the proof which contains the most cuts.*

In any case, this is a form of abstraction. Which is no surprise since we already noticed that Kolmogorov complexity is fundamentally related to the notion of abstraction. Thus one can say that: *Knowledge is abstract information: abstract, compressed, with some intensionality content.*

And such a knowledge will be, in its turn, compressed, etc. This is exactly the mode the brain functions with language and mathematics. Observe that some abstractions are somewhat "accidental": they occurred at some time and drastically modify the state of knowledge. Note that it is really what Kolmogorov complexity shows. Suppose an integer has a long and seemingly lawless binary (or decimal) representation: it takes space to represent it in this way. But if we get a (good) constructible property about this integer, then we can obtain a short, abstract, compressed characterization of it. And this increases our knowledge. In the same way, the development of integral calculus, some parts of geometry and fractal geometry, allow for short (effectively computable) sequential descriptions of shapes. Especially, it appears that Kolmogorov's complexity can be a very useful theory in order to address in a mathematical way the approaches of classification, which are now essentially, to the exception relational database, heuristic methods (not yet fully formalized as can be expected from a classification method) such as classification using compression and Google classification. One can also hope for applications in other domains such as semiology, cognitive science or biology with the genome, as spectacularly shown by the French biologist Antoine Danchin in his book (Danchin 1998). Indeed, classification by compression is already used by some biologists in such a perspective.

Let us conclude by stressing again how much useful are such classification methods using compression or using Google along the top-down operational mode. In many cases, we face huge families of objects (when one can define them) for which there is no obvious structure. *So that we really are in a syntactic world and want to grasp this world with some semantic.* This is, for example, the case for DNA sequences of living organisms and for the multi billion many files on the Web...

For that last example, though we are not so much pessimistic, let us cite Edsger W. Dijkstra's penetrating analysis in his famous 1972 Turing award reception speech (Dijkstra 1972)[52]:

> As long as there were no machines, programming was no problem at all; when we had a few weak computers, programming became a mild problem, and now that we have gigantic computers, programming has become an equally gigantic problem. In this sense the electronic industry has not solved a single problem, it has only created them – it has created the problem of using its products.

Acknowledgements For Francine Ptakhine, who gave me liberty of thinking and writing.

Thanks to Serge Grigorieff and Chloé Ferbus for listening, fruitful communication and for the careful proofreading and thanks to Maurice Nivat who welcomed me at the LITP in 1983.

References

Bennett, C. (1988). Logical depth and physical complexity. In R. Herken (Ed.), *In the universal turing machine: A half-century survey* (pp. 227–257). Oxford University Press: New York.

Bennett, C., Gács, P., Li, M., Vitányi, P., & Zurek, W. (1998). Information distance. *IEEE Transactions on Information Theory, 44*(4), 1407–1423.

Chaitin, G. (1966). On the length of programs for computing finite binary sequences. *Journal of the ACM, 13*, 547–569.

Chaitin, G. (1969). On the length of programs for computing finite binary sequences: Statistical considerations. *Journal of the ACM, 16*, 145–159.

Chaitin, G. (1975). A theory of program size formally identical to information theory. *Journal of the ACM, 22*, 329–340.

Chen, P. S. (1976). The entity-relationship model: Toward a unified view of data. *ACM Transactions on Database Systems, 1*(1), 9–36.

Cilibrasi, R. (2003). Clustering by compression. *IEEE Transactions on Information Theory, 51*(4), 1523–1545.

Cilibrasi, R., & Vitányi, P. (2005). Google teaches computers the meaning of words. *ERCIM News, 61*.

Cilibrasi, R., & Vitányi, P. (2007). The Google similarity distance. *IEEE Transactions on Knowledge and Data Engineering, 19*(3), 370–383.

Codd, E. W. (1970). A relational model of data for large shared databanks. *Communications of the ACM, 13*(6), 377–387.

Codd, E. W. (1990). *The relational model for database management. Version 2*. Reading: Addison-Wesley.

Danchin, A. (1998). *The delphic boat: What genomes tell us*. Paris: Odile Jacob [Cambridge/London: Harvard University Press, 2003.]

[52] Let us mention the remarkable collection of unpublished papers and notes (Dijkstra 1982).

Delahaye, J. P. (1999). *Information, complexité, hasard* (2nd ed.). Hermès.
Delahaye, J. P. (2004). Classer musiques, langues, images, textes et génomes. *Pour La Science, 316*, 98–103.
Delahaye, J. P. (2006). *Complexités: Aux limites des mathématiques et de l'informatique*. Belin-Pour la Science.
Dershowitz, N., & Gurevich, Y. (2008). A natural axiomatization of computability and proof of Church's thesis. *The Bulletin of Symbolic Logic, 14*(3), 299–350.
Dijkstra, E. W. (1972). The humble programmer. In *ACM Turing Lecture*. Available on the Web from http://www.cs.utexas.edu/~EWD/transcriptions/EWD03xx/EWD340.html
Dijkstra, E. W. (1982). *Selected writings on computing: A personal perspective*. New York: Springer.
Durand, B., & Zvonkin, A. (2004–2007). Kolmogorov complexity, In E. Charpentier, A. Lesne, & N. Nikolski (Eds.), *Kolmogorov's heritage in mathematics* (pp. 269–287). Berlin: Springer. (pp. 281–300, 2007).
Evangelista, A., & Kjos-Hanssen, B. (2006). Google distance between words. *Frontiers in Undergraduate Research*, University of Connecticut.
Feller, W. (1968). *Introduction to probability theory and its applications* (3rd ed., Vol. 1). New York: Wiley.
Ferbus-Zanda, M. (1986). *La méthode de résolution et le langage Prolog* [*The resolution method and the language Prolog*], Rapport LITP, No-8676.
Ferbus-Zanda, M. (In preparation-a). *Logic and information system: Relational and conceptual databases*.
Ferbus-Zanda, M. (In preparation-b). *Kolmogorov complexity and abstract states machines: The relational point of view*.
Ferbus-Zanda, M. (In preparation-c). *Duality: Logic, computer science and Boolean algebras*.
Ferbus-Zanda, M. (In preparation-d). *Logic and information system: Cybernetics, cognition theory and psychoanalysis*.
Ferbus-Zanda, M., & Grigorieff, S. (2004). Is randomness native to computer science? In G. Paun, G. Rozenberg, & A. Salomaa (Eds.), *Current trends in theoretical computer science* (pp. 141–179). River Edge: World Scientific.
Ferbus-Zanda, M., & Grigorieff, S. (2006). Kolmogorov complexity and set theoretical representations of integers. *Mathematical Logic Quarterly, 52*(4), 381–409.
Ferbus-Zanda, M., & Grigorieff, S. (2010). ASMs and operational algorithmic completeness of lambda calculus. In N. Dershowitz & W. Reisig (Eds.), *Fields of logic and computation* (Lecture notes in computer science, Vol. 6300, pp. 301–327). Berlin/Heidelberg: Springer.
Howard, W. (1980). The formulas-as-types notion of construction. In J. P. Seldin & J. R. Hindley (Eds.), *Essays on combinatory logic, lambda calculus and formalism* (pp. 479–490). London: Academic.
Irving, J. (1978). *The world according to garp*. Modern Library. Ballantine.
Kolmogorov, A. N. (1956). *Grundbegriffe der Wahscheinlichkeitsrechnung* [Foundations of the theory of probability]. New York: Chelsea Publishing. (Springer, 1933).
Kolmogorov, A. N. (1965). Three approaches to the quantitative definition of information. *Problems of Information Transmission, 1*(1), 1–7.
Kolmogorov, A. N. (1983). Combinatorial foundation of information theory and the calculus of probability. *Russian Mathematical Surveys, 38*(4), 29–40.
Krivine, J. L. (2003). Dependent choice, 'quote' and the clock. *Theoretical Computer Science, 308*, 259–276. See also http://www.pps.univ-paris-diderot.fr/~krivine/
Li, M., & Vitányi, P. (1997). *An introduction to Kolmogorov complexity and its applications* (2nd ed.). New York: Springer.
Li, M., Chen, X., Li, X., Mav, B., & Vitányi, P. (2003). The similarity metrics. In *14th ACM-SIAM 1357 symposium on discrete algorithms*, Baltimore. Philadelphia: SIAM
Martin-Löf, P. (1979, August 22–29). *Constructive mathematics and computer programming*. Paper read at the 6-th International Congress for Logic, Methodology and Philosophy of Science, Hannover.

Shannon, C. E. (1948). The mathematical theory of communication. *Bell System Technical Journal, 27*, 379–423.

Shoenfield, J. (2001). *Recursion theory*, (Lectures notes in logic 1, New ed.). Natick: A.K. Peters

Solomonoff, R. (1964). A formal theory of inductive inference, Part I & II. *Information and control, 7*, 1–22; 224–254.

Véronis, J. (2005). Web: Google perd la boole. (Web: Googlean logic) Blog. January 19, 2005 from http://aixtal.blogspot.com/2005/01/web-google-perd-la-boole.html, http://aixtal.blogspot.com/2005/02/web-le-mystre-des-pages-manquantes-de.html, http://aixtal.blogspot.com/2005/03/google-5-milliards-de-sont-partis-en.html.

Wiener, N. (1948). *Cybernetics or control and communication in the animal and the machine* (2nd ed.). Paris/Hermann: The Technology Press. (Cambridge: MITs, 1965).

Chapter 5
Proof-Theoretic Semantics and Feasibility

Jean Fichot

5.1 Introduction

It is well known that classical mathematics have been widely criticised by the proponents of constructive mathematics on the ground that the former rests on a realist conception of the realm of mathematical objects. It must be recalled that the latter is not completely immune to such a reproach inasmuch as some vestige of realism is still present in its foundations (1). This realism takes the form of two different idealisations of human abilities: the creative ones and the mechanical ones (1.1). It can be argued that the first idealisation is avoided by the proof theoretic-semantics of constructive mathematics but not the second one (1.2). Different definitions of feasible functions and systems of feasible mathematics have been proposed that makes possible to avoid the second idealisation too (2). It is of special interest to see if they allow, at least partially, some proof-theoretical semantics (2.2).

5.2 Idealised Foundations of Mathematics

The fact that constructive mathematics rests on a strong idealisation of the abilities of flesh and blood mathematicians is well known. Its origin is to be found in Borel who remarked that there is an analogy between the passage from finite numbers to huge finite numbers and the passage from natural numbers to infinite ordinals (Borel 1950). Borel's idea was used by van Dantzig in an argument that was supposed to shed some doubts on the possibility to draw a sharp distinction between "formalist"

J. Fichot (✉)
Université Paris 1 Panthéon-Sorbonne – IHPST, Paris, France
e-mail: Jean.Fichot@univ-paris1.fr

J. Dubucs, M. Bourdeau (eds.), *Constructivity and Computability in Historical and Philosophical Perspective*, Logic, Epistemology, and the Unity of Science 34,
DOI 10.1007/978-94-017-9217-2_5

and "intuitionistic" foundations of mathematics as far as the one between finite numbers and infinite ones is also gradual (van Dantzig 1956). A number like $10^{10^{10}}$ may be seen as a finite number but also as a number that has never been and will never be actually constructed; as such this number might be compared with a purely "formal" number like $\omega^{\omega^{\omega}}$. From the fact that intuitionism does not distinguish between what can be actually constructed and constructions that we can only imagine to be performed, van Dantzig concluded that it was at least partially formal. Of course, most of the proponents of constructive mathematics were quite aware of the necessity of this idealisation and it is instructive to compare the ways Brouwer and Bishop viewed it.

5.2.1 Two Different Kinds of Idealisation

Among a mathematician's abilities a distinction must be made between the creative ones and the purely mechanical ones. Mechanical abilities depend only on resources of memory, time, space. For example, the only idealisation involved in Turing's analysis is the idealisation of mechanical abilities since no specific creative abilities are allowed to the (human) computer (Turing 1936–37). And obviously the concept of creating, or creative, subject, also known as the idealised mathematician, introduced by Brouwer in (1948), involves an idealisation of both aspects. This foundation of constructive mathematics on the metaphysical hypothesis of the creative subject is in fact very problematic because, as it is emphasized by van Dalen and Troelstra, it entails that there are two different kind of intuitionistic mathematics: the idealised mathematician's one and the one that is practised by the flesh and blood mathematicians with strictly finite cognitive powers (Troelstra and van Dalen 1988). Mathematical certainty belongs only to the former, not to the later since it depends on language which is a limited and fallible mean to communicate and remember mathematical constructions.

> [...] it may well be that we have no proper understanding what mathematics according to Brouwer's principles really looks like, since the divergence between mathematics conforming to the theoretical postulate of languageless mental activity and our actual (intuitionistic) practice is too great. (832)

Now, if we turn to Bishop some idealisation is indeed assumed as a proviso on which the mere possibility of constructivisation of classical analysis depends (Bishop 1967).

> The transcendance of mathematics demands that it should not be confined to computations that I can perform, or you can perform, or 100 men working 100 years with 100 digital computers can perform. Any computation that can be performed by a finite intelligence - any computation that has a finite number of steps - is permissible. (3)

One point must be emphasized: in spite of the mention of "*a finite intelligence*" in this quotation, Bishop seems to refer to some intuitive notion of mechanical computation and the idealisation here is only the one of mechanical abilities, not of creative abilities. But this is not the main difference with Brouwer's creative subject; Bishop in fact considered the constructivisation of classical mathematics as a first necessary step in a much more ambitious programme of "efficient" constructivisation of mathematics.

> This does not mean that no value is to be placed on the efficiency of a computation. An applied mathematician will prize a computation for its efficiency above all else, whereas in formal mathematics much attention is paid to elegance and little for efficiency. Mathematics should and must concern itself with efficiency, perhaps to the detriment of elegance, but the matters will come to the fore only when realism has begun to prevail. Until then our first concern will be to put as much mathematics on a realistic basis without close attention to questions of efficiency. (3)

5.2.2 *Proof-Theoretic Semantics and Idealisation*

It could be argued that the notion of creative subject is not the only possible way to give a philosophical foundation of constructive mathematics. Another path has been explored which supposes to give a meaning theoretical account of the language of constructive mathematics. This idea has a long history that began with Gentzen's proof theoretical investigations and the informal explanation of the constructive meaning of the logical constants as given by Brouwer, Heyting and Kolmogorov, the BHK interpretation.

I will not enter here into the details of this story; suffices it to say that the main problem was to give a theoretical analysis of the relation between two notions of proof: the one that was used in the BHK interpretation and the one of real proofs. To put it in a nutshell, the problem arises because the former, now known as the notion of canonical proof, is indeed rather strange – for example a canonical proof of a disjunction is a proof of one of its members – when compared to the later; who has ever proved in practice a disjunction this way? As it has often be emphasized, the concept of proof used in the BHK interpretation is not immediately closed under *modus ponens* since $(B \vee C)$ may be deduced from proofs of $A \Rightarrow (B \vee C)$ and A and this does not tell which one of B and C is true.

Heyting's solution to this problem was given in successive steps culminating in the idea of proof as function: a proof of $A \Rightarrow B$ should be a function that, applied to any canonical proof of A yields a canonical proof of B. Since real proofs are of course non canonical proofs, this was supposed to explain the relationship between them: a real proof of A must be thought as a potential construction that, when completely performed, should yield canonical proof of A. Once this distinction between potential and actual construction was generalised, the problem that remained was to explain how the process of actualisation can be performed. Indeed this idea of proof as function, as well accepted as it is nowadays, looked in fact completely mysterious for quite a long time since it supposes to identify proofs, even formal ones, with computable functions. But the former seemed to belong to the word of static, inert, objects and the later rather to the one of dynamic processes and those two worlds were not easily thought of as overlapping.

Let us give two examples. The first functional interpretation of Heyting arithmetic was given by Kleene who later admitted frankly that at that time the BHK interpretation was of no help for him; the choice made by Kleene to deal only with recursive functions was a strong obstacle that prevented him to give a satisfactory analysis of the dual concept of proof-as-function (Kleene 1973). Nearly at the same time, Gödel was led to his own interpretation of Heyting arithmetic by functionals of finite type, the *Dialectica* interpretation, because he doubted that proofs in this formal system were indeed constructive according to the criteria given by the BHK interpretation. It was only around 1970 that he came to see these computable functionals of finite type as proofs albeit given in another system (Gödel 1958, 1990). At that moment, the link between cut elimination in proofs and formal computation of terms had already been made by the Curry-Howard correspondence (Curry and Feys 1958; Howard 1980) on which proof-theoretic semantics, strongly advocated by Dummett, Prawitz, Martin-Löf and other authors, rests (Dummett 1975; Gentzen 1935; Howard 1980; Martin-Löf 1984, 1985; Prawitz 1965).

Formal proofs are conceived, not only as static objects, but also as dynamical processes. According to this analysis, the meaning of a logical or mathematical constant is given by two kinds of rules: the ones that are used in the proof, introduction and elimination rules for that constant, and computation rules, reduction and expansion, that operate on the proofs and justify the former.

- The reduction rules show that the elimination rule is not *stronger* then the introduction rule. From a computational point of view, they are indeed the most important rules. From a meaning theoretical point of view the association of the introduction/elimination rules and the reduction rules was used as an argument by Prawitz and Dummett that was supposed to show why classical logic is partially meaningless since the elimination rule for $\vee$ seems, at least from a constructive point of view, stronger than its classical introduction rules.

$$\dfrac{\dfrac{\begin{matrix}\vdots\\ \Gamma, x : A \vdash b : B\end{matrix}}{\Gamma \vdash \lambda x.b : A \Rightarrow B}{\scriptstyle(\Rightarrow i)} \quad \begin{matrix}\vdots\\ a : A\end{matrix}}{\Gamma \vdash (\lambda x.b)\, a : B}{\scriptstyle(\Rightarrow e)} \overset{\beta}{\twoheadrightarrow} \begin{matrix}\vdots\\ \Gamma \vdash b\,[a/x].b\end{matrix}$$

- The expansion rules have only a very weak computational content; in the case of implication, read from right to left, the expansion rule yields an η-reduction rule: $\lambda x\,(t)\,x \overset{\eta}{\twoheadrightarrow} t$, if x is not free in t. From a meaning theoretic point of view, those rules show that the introduction rule is not *stronger* than the elimination rule.

$$\begin{matrix}\vdots\\ \Gamma \vdash t : A \Rightarrow B\end{matrix} \overset{\eta}{\rightrightarrows} \dfrac{\dfrac{\begin{matrix}\vdots\\ \Gamma \vdash t : A \Rightarrow B\end{matrix} \quad x : A \vdash x : A}{\Gamma, x : A \vdash (t)\, x : B}{\scriptstyle(\Rightarrow e)}}{\Gamma \vdash \lambda x.\,(t)\, x : A \Rightarrow B}{\scriptstyle(\Rightarrow i)}$$

The $\beta\eta$ reduction rules give a mathematical content to the concept of proof-as-function since they make it possible to see a formal deduction of a formula as an algorithm—a lambda term—that yields in a finite number of steps a unique value; a deduction Π of a sequent $A_1, \ldots, A_n \vdash B$ is an algorithm that yields, for any deduction $\Pi_1, \ldots, \Pi_n$ of the $A_1, \ldots, A_n$, a deduction of A. And the latter is a cut-free deduction that ends with a step of introduction for the main constant of A; in the case of a logical constant this deduction satisfies the criterion of the BHK interpretation and may be held as a formal, mathematically defined, version of the intuitive concept of canonical proof.

BHK interpretation intuitive notion	Proof theoretic semantics formal definition
Real proofs potential constructions	Indirect deductions
Canonical proofs actual constructions	Cut-free deductions

As shown in the table above, this is indeed a very satisfactory theoretical account of the constructive meaning of the logical and mathematical constants as given by the BHK interpretation. Notice also that the idealisation of creative abilities is not explicitly used in proof-theoretic semantics. The only problem is that it does not completely avoid the issue of idealised mechanical abilities; in fact it can help to give it a precise content.

β-reduction and μ-expansion give only a local, and as such partial, justification of the introduction and elimination rules for a given constant. For example, it is easy to check the so-called Prawitz's rules for $\in$

$$\frac{\Gamma \vdash A\,[t/y]}{\Gamma \vdash t \in \{y; A\}}\,(\in i) \qquad \frac{\Gamma \vdash t \in \{y; A\}}{\Gamma \vdash A\,[t/y]}\,\in e$$

are locally justified. But in the system obtained by adding to them the elimination and introduction rules for implication there is a "proof" of $\vdash (\upsilon)\,\upsilon : C$, where $\upsilon \equiv \lambda x\,(x)\,xx$ and C is an arbitrarily chosen formula. This implies that a system of rules must be globally justified by a cut elimination theorem and this may involve a new proof when some constants are added to a given system since the union of two globally justified systems may not be globally justifiable.[1]

Since a cut elimination theorem must be given by a (real) indirect proof according to some rules that cannot be part of the ones it justified, the search for a global

[1]Notice also that this is a very strong argument that can be used against a molecularist conception of meaning according to which the meaning of a constant is given only by its introduction and elimination rules: in fact the global justification of these rules may depend on the rules governing the use of the other constants. For example, the system obtained by adding to Prawitz's rules for $\in$ the rules for linear implication is globally justified.

justification is obviously open ended.[2] Moreover, in the case of, say, Heyting Arithmetic, this theorem entails its consistency. But for some Π_1^0 formulas $\forall x A$ independent of Heyting Arithmetic and provable in some higher order system, the length of the cut-free proofs of $A\,[n/x]$ is unbounded, more precisely, for each $m \in \mathbb{N}$, there is a $n \in \mathbb{N}$ such that the length of the cut-free proof of $A\,[n/x]$ in Heyting Arithmetic is greater than m (Buss 1987; Gödel 1936). And even for formulas already provable in intuitionistic logic, it is well known that cut free proofs may need to be superexponentially larger than proofs that contain cut. According to a physical bound, like Beckenstein's one, this implies that canonical proofs are ideal abstract objects. Indeed, there is something strange in the use of the words "actual" and "potential" since the former qualifies constructions that will never be actualized and the latter qualifies real proofs. In fact, canonical proofs in the BHK interpretation could rather be thought of as a refinement of the notion of truth; compare "$A_1 \vee A_2$ is true if and only if A_1 is true or A_2 is true" with "$\langle p, i\rangle$ is a proof of $A_1 \vee A_2$ if and only if p is a proof of A_i". One may wonder what such a notion of truth has to do with the much more plausible idea that a proposition is true if it has been proved by a real proof.

Now here is the main problem we face with Dummett and Prawitz's philosophical foundations of mathematics: how can we explain the meaning of their language, as it is used in actual proofs, which are obviously length-feasible, by means of canonical proofs, which most of the time are not length-feasible? If we think, metaphorically, of a proof as a path that leads us to the experience of truth, then it should be clear that it is not a shortcut climbing up the hill that we take just to avoid the longer path in the valley that leads to the same destination, because there is no longer path there that we could take; of course a pedestrian can always imagine that there is some secret road that could have been taken by some idealised traveller (keep on straight ahead for $10^{10^{10}}$ miles!) but it would be of little help to him.

There is one possible way to avoid this discrepancy between proof theoretical meaning explanations and non feasible canonical proofs. It involves some lightening of the logical and mathematical rules that allow to avoid non feasible proofs; but then of course the problem we have to face here is to show that those rules can, at least partially, give some proof-theoretic semantics.

5.3 Feasible Mathematics and Meaning Explanations

There is a strong temptation here to paraphrase and strengthen Bishop's dictum: *Mathematics belongs to man, not to the creative subject* [...] *If the creative subject has mathematics of his own that needs to be done, let him do it himself.* If we

[2]It could be argued that this shows that some idealisation of the creative abilities is still present in proof-theoretic meaning explanations at least if we think that such explanations should be definitely consistent. But there are good reasons, which can be found in Martin-Löf (1987), to believe that this hope rests on a confusion between validity and truth.

give in to this temptation, then we have to face seriously Bishop's idea that the constructivisation of classical mathematics is only a first step and that the next one should be an efficient constructivisation of mathematics. Since nowadays systems of feasible computations and proofs exist that can give some content to the idea that even canonical proofs should be feasible, it is of special interest to see if they allow for some meaning theoretical account in the tradition of Dummett and Prawitz.

Wright, Dubucs, Marion and other authors have emphasized the importance of the distinction between feasible constructions and the ones that could be carried out only in principle (Dubucs 1997; Dubucs and Marion 2003; Wright 1982). Different attempts have been made to extend this distinction amongst finite objects—constructions, proofs, computable functions—between the admissible or feasible ones and the idealised ones. One point worth mentioning is that, even in the case of the natural numbers, the main difficulty is to draw a sharp line between the former and the latter since the fuzzy concept of feasible number involves a well known sorites paradox. First, it is obvious that the successor operation $x \mapsto x + 1$ is computation-feasible since it involves a strictly finite number of steps of computation depending of course on the chosen formal representation of the numbers, but not on the length of the representation of x. Second, it must be taken for granted that most natural numbers are not admissible or feasible. But then we come to a paradox: starting from an admissible number, we can feasibly compute $x + 1$ and 0 is indeed admissible; then what conclusion should we come to? That all natural numbers are feasible? This is obviously false, but as we shall see, there is another possible answer which is to hold the distinction between feasible and non-feasible numbers to be relative.

5.3.1 Two Implicit Definitions of Feasible Functions

The identification of the intuitive notion of feasible function with that of function computable in number of steps given by a polynomial function of the length of its inputs is generally admitted on the basis of arguments given by Cobham in (1965). The first implicit definition of this class of functions rests on a distinction between the set of feasible numbers and the set of natural numbers which was given by Bellantoni and Cook in (1997).[3] Leivant gave his own definition which can be seen

[3]Of course, a function $f : \mathbb{N} \to \mathbb{N}$ is polytime if there is some Turing machine $\mathcal{M}$ that computes the value of $f(n)$ in $P(|n|)$ steps, for a polynomial function P of the length $|n|$ of n. But this is an explicit definition of polytime functions, not an implicit one. Bellantoni and Cook's definition is not the first implicit definition of the class of polytime functions. In Cobham (1965), definitions such as $f(x + 1, y) = g(x, y, f(x, y))$ are admitted as far as for some previously defined function h we have $f(x, y) \leq h(x, y)$, for all x, y. An ad hoc primitive function must be given $x, y \mapsto 2^{|x|.|y|}$.

as typed version of Bellantoni and Cook's definition, and may help to understand the latter (Leivant 1993). In both cases, the main idea is that a definition like

$$F = \{y \in \mathbb{N} / \exists x \in \mathbb{N} f(x) = y\}$$

when $f(x+1) = g(f(x))$ where g is some previously defined function, involves some impredicativity since the existence of y' such that $y' = f(x+1) = g(f(x))$ depends on a recursive call on the unbounded value of $f(x) > x+1$.

In order to avoid such definitions, Leivant introduces two sets of natural numbers, $\mathbb{N}_0$ for the feasible numbers and $\mathbb{N}_1$ for the natural numbers. Despite the fact that these two sets are isomorphic copies of $\mathbb{N}$, the definition of an element of $\mathbb{N}_0$ is not allowed to depend on the existence of elements of $\mathbb{N}_1$ which contains all the numbers that are defined by use of elements of $\mathbb{N}_0$. According to Cook and Bellantoni, the arguments in a function may appear in normal positions for feasible numbers and safe ones for natural numbers, those positions being separated by a semicolon. Then a function f, is such that

$$f(\overbrace{x_1, \ldots x_n}^{\text{normal}}; \overbrace{y_1, \ldots y_m}^{\text{safe}}), \text{ is of type } \overbrace{\mathbb{N}_0, \ldots, \mathbb{N}_0}^{n}, \overbrace{\mathbb{N}_1, \ldots, \mathbb{N}_1}^{m} \to \mathbb{N}_1$$

The class of feasible functions, LBC, contains a stock of primitive functions, constant functions, successor, projections, predecessor.[4] During a computation, it should not be possible to recurse on a safe input, i.e. a non feasible number in $\mathbb{N}_1$, only on a normal input, i.e. a feasible one in $\mathbb{N}_0$. According to this restriction, the safe recursion scheme is:

- If $g : \mathbb{N}_0^n, \mathbb{N}_1^m \to \mathbb{N}_1$,$h : \mathbb{N}_0^{n+1}, \mathbb{N}_1^{m+1} \to \mathbb{N}_1$ are in LBC, then $f : \mathbb{N}_0^{n+1}, \mathbb{N}_1^m \to \mathbb{N}_1$ defined by

$$\begin{cases} f\left(0, x_1^n; y_1^m\right) = g\left(x_1^n; y_1^m\right) \\ f\left(x+1, x_1^n; y_1^m\right) = h\left(x, x_1^n; f\left(x, x_1^n; y_1^m\right), y_1^m\right) \end{cases}$$

is in LBC. $\left(x_1^n \equiv x_1, \ldots x_n, y_1^m \equiv y_1, \ldots y_m\right)$.

For example, since recursion is allowed only on $\mathbb{N}_0$, for $h : \mathbb{N}_0, \mathbb{N}_1 \to \mathbb{N}_1$, f defined by $f(Sx; y) = h(x; f(x, y))$ is of type $\mathbb{N}_0, \mathbb{N}_1 \to \mathbb{N}_1$. Thus, the set $F = \{z / \exists x \in \mathbb{N}_0 \exists y \in \mathbb{N}_1 f(x; y) = z\}$ is included in $\mathbb{N}_1$ but not in $\mathbb{N}_0$.

The composition scheme is also modified in such a way that it is impossible for an argument to cross the semicolon from left to right, i.e. by defining a function g such that some arguments in a normal position in the definition of a function f

---

[4]In order to save space, we use all through this paper a unary representation of natural numbers. But it should be clear that we should have chosen to deal with binary words built from a constant for the empty word and two successors.

already in LBC appear in a safe position in g. For example, if $f(x; y)$ is in BC, $g(; x, y) = f\left(\pi_1^2(; x, y); \pi_2^2(; x, y)\right)$ which could give $g(; x, y) = f(x; y)$, is not an admissible equation.

5.3.1.1 Two Theorems

Polytime soundness *if f is in LBC, then f is polytime computable.*
Polytime completeness *if g is polytime computable, then there is a function f in LBC such that $f =_{ext} g$.*

Three important remarks must be made.

1. The soundness theorem above holds only if the value of $f(t_1, \ldots t_k; u_1, \ldots u_l)$ is computed by call by value, i.e. if the values n_i, m_j of the terms t_i, u_j are computed before the one of $f(n_1, \ldots n_k; m_1, \ldots m_l)$.
2. In the completeness theorem the equality in $f =_{ext} g$ is extensionnal: f and g yields the same value for the same arguments but in many cases the algorithm that computes the values of g is not faithfully translated in the equations system that defines f.
3. Despite the preceding remarks, the fascinating insight due to Bellantoni's, Cook's and Leivant's definitions is that the distinction amongst natural numbers between the feasible ones and the non feasible ones is not absolute but relative to the computation that is performed.

The definition of LBC has been extended into a typed system that allows for definition of functions by restricted higher type recursion schemes in Hofmann (2000). Later on, this system has been used to give a realisability interpretation of an axiomatisation of feasible arithmetic in Bellantoni and Hofmann (2002).

5.3.2 *A Feasible Arithmetic*

As given by Bellantoni and Hofmann, feasible arithmetic *FA* is an axiomatised system very close to Peano Arithmetic with a modal operator $\Box$ that allows the contraction of hypotheses in a derivation. In order to comply with the criteria of proof-theoretic semantics, we introduce the main rules of this system in a natural deduction style. Our main inspiration here is to be found in Davies and Pfenning (2001).

5.3.2.1 Sequent

The left hand side of a sequent is split in two different areas separated by a double stroke. The left area contains formulas that are implicitly modal and express perennially, or eternally, true propositions, the right area contains formulas that express simply true propositions.

$$\overbrace{G_1, \ldots, G_m}^{\text{perennial}} \parallel \overbrace{D_1, \ldots, D_m}^{\text{true}} \vdash C \text{ true}$$

Notation if $\Lambda_i = \{A_1, \ldots, A_m\}$ and $\Lambda_j = \{B_1, \ldots, B_n\}$ are multisets, then $\Lambda_{i,j}$ denotes the multiset $\{A_1, \ldots, A_m, B_1, \ldots, B_n\}$.

5.3.2.2 Feasible Logic

Axioms

There are two kinds of axioms. The first ones are just identity axioms corresponding to $A \supset A$, but the second ones have a modal content since they correspond to $\Box A \supset A$ and express the fact that an eternally true proposition is true.

$$\overset{axiom}{.\| A \vdash A} \qquad \overset{axiom}{A \,\|. \vdash A}$$

Logical Rules

The logical rules are the ones of natural deduction. Notice that only true formulas are active.

$$\frac{\Gamma \,\|\Delta, A \vdash B}{\Gamma \,\|\Delta \vdash A \supset B}(\supset i) \qquad \frac{\Gamma_0 \,\|\Delta_0 \vdash A \supset B \quad \Gamma_1 \,\|\Delta_1 \vdash A}{\Gamma_{0,1} \,\|\Delta_{0,1} \vdash B}(\supset e)$$

$$\frac{\Gamma_1 \,\|\Delta_1 \vdash A \quad \Gamma_2 \,\|\Delta_2 \vdash B}{\Gamma_{1,2} \,\|\Delta_{1,2} \vdash A \wedge B}(\wedge i)$$

$$\frac{\Gamma_1 \,\|\Delta_1 \vdash A \wedge B \quad \Gamma_2 \,\|\Delta_2, A, B \vdash C}{\Gamma_{1,2} \,\|\Delta_{1,2} \vdash C}(\wedge e)$$

$$\left\{\frac{\Gamma \,\|\Delta \vdash A_i}{\Gamma \,\|\Delta \vdash A_1 \vee A_2}(\vee i)\right\}^{1 \leqslant i < 2} \qquad \frac{\Gamma_0 \,\|\Delta_0 \vdash A_1 \vee A_2 \quad \{\Gamma_i \,\|\Delta_i, A_i \vdash C\}^{1 \leqslant i < 2}}{\Gamma_{0,1,2} \,\|\Delta_{0,1,2} \vdash C}(\vee e)$$

Modal Rules

The rules of introduction and elimination rules for $\Box$.

$$\frac{\Gamma \,\|. \vdash A}{\Gamma \,\|. \vdash \Box A}(\Box i) \qquad \frac{\Gamma_0 \,\|\Delta_0 \vdash \Box A \quad \Gamma_1, A \,\|\Delta_1 \vdash C}{\Gamma_{0,1} \,\|\Delta_{0,1} \vdash C}(\Box e)$$

The intuitive justifications of both are to be found in the idea that the formulas on the right hand side of the double stroke are eternally true propositions. For example, A in the antecedent of the right premiss of $(\Box e)$ is equivalent to $\Box A$ in the left premiss. Indeed, with those rules, "A is eternally true" is internalised by "$\Box A$ is true" since the rules

$$\downarrow \frac{\Gamma, A \parallel \Delta \vdash C}{\Gamma \parallel \Box A, \Delta \vdash C} \uparrow$$

are derivable.

Structural Rules

The rules of contraction and weakening.

$$\frac{\Gamma, A, A \parallel \Delta \vdash C}{\Gamma, A \parallel \Delta \vdash C}\,(contr) \qquad \frac{\Gamma \parallel \Delta \vdash C}{\Gamma, A \parallel \Delta, B \vdash C}\,(weak)$$

Contraction is allowed only for formulas A eternally true or, equivalently by internalisation, for formulas $\Box A$ simply true.

Justification of the Rules

The justification of the introduction and elimination rules for each logical constants can be given by reduction and expansion rules that operate on proofs. For example, in the case of $\Box$, we have

- $$\dfrac{\dfrac{\begin{array}{c}\mathcal{D}_0\\ \Gamma_0 \parallel . \vdash A\end{array}}{\Gamma_0 \parallel . \vdash \Box A}\,(\Box i) \qquad \begin{array}{c} A \parallel . \vdash A \\ \mathcal{D}_1 \\ \Gamma_1, A \parallel \Delta_1 \vdash C\end{array}}{\Gamma_{0,1} \parallel \Delta_1 \vdash C}\,(cut) \rightarrow_\beta \begin{array}{c}\mathcal{D}_0\\ \Gamma_0 \parallel . \vdash A \\ \mathcal{D}_1 \\ \Gamma_{0,1} \parallel \Delta_1 \vdash C\end{array}$$
- $$\begin{array}{c}\vdots\\ \Gamma \parallel \Delta \vdash \Box A\end{array} \Rightarrow_\eta \dfrac{\begin{array}{c}\vdots\\ \Gamma \parallel \Delta \vdash \Box A\end{array} \quad (\Box i)\dfrac{A \parallel . \vdash A}{A \parallel . \vdash \Box A}}{\Gamma \parallel \Delta \vdash \Box A}\,(\Box e)$$

Two points are worth mentioning. First, as emphasised by Bellantoni and Hoffmann, feasible logic is strongly reminiscent of the modal logic S4 used by Shapiro in his system of epistemic arithmetic (Shapiro 1985), and, indeed, when formulated in a natural deduction style, the rules of feasible logic are nearly the same as the ones given by Davies and Pfenning for their constructive version of S4 (Davies and Pfenning 2001); for example $\vdash \Box A \supset (A \wedge \Box\Box A)$ is probable. The only difference with the latter is the use of $\supset$ and $\Box$ which allows contraction

$A \supset (A \wedge A)$ only if A is $\Box B$ for some formula B. Second, if we replace in those rules the logical constants $\supset, \wedge, \vee$ by $\multimap, \otimes, \oplus$, and the modal operator $\Box$ by !, we obtained the rules of intuitionistic linear logic extended with the weakening rule, known as intuitionist affine logic in the linear logic community. Indeed these are the notations that are actually used in Hofmann's realisability interpretation of feasible arithmetic. The fact that the rules of feasible, or affine linear, logic can be justified by local expansion and reduction rules, could be taken to be a rather encouraging sign that some theoretic semantics of feasible logic could be given. Alas, we shall see this hope cannot be fulfilled, but we have to postpone this discussion for a moment.

Natural Numbers and Induction Principle

The axioms for S (successor) are given by introduction rules for a predicate constant **N**

$$\frac{axiom}{.\,\|.\ \vdash \mathbf{N}0} \qquad\qquad \frac{\Gamma \,\|\Delta\ \vdash \mathbf{N}t}{\Gamma \,\|\Delta\ \vdash \mathbf{N}St}$$

With those rules and the ones for $\Box$, the derivations of the introduction rules for the complex predicate $\Box\mathbf{N}$ are immediate. Contraction is also allowed for **N**

$$\frac{\Gamma \,\|\Delta, \mathbf{N}t, \mathbf{N}t \vdash C}{\Gamma \,\|\Delta, \mathbf{N}t \vdash C}\,(contr)$$

There is no induction axiom schema for **N**, but only a feasible induction schema for $\Box\mathbf{N}$

$$A\,[0/x] \supset [\Box\forall y\,[\Box\mathbf{N}y \supset (A\,[y/x] \supset A\,[Sy/x])] \supset \forall x\,(\Box\mathbf{N}x \supset A)]$$

where A does not contain $\Box$.

Feasible induction can be formulated equivalently as an elimination rule for $\Box\mathbf{N}$

$$\frac{\begin{array}{c}\mathcal{D}_0\\ \Gamma_0 \,\|\Delta_0 \vdash A\,[0/x]\end{array} \quad \begin{array}{c}.\,\|A\,[y/x] \vdash A\,[y/x] \quad .\,\|\Box\mathbf{N}y \vdash \Box\mathbf{N}y\\ \mathcal{D}_1\\ \Gamma_1 \,\|\Box\mathbf{N}y, A\,[y/x] \vdash A\,[Sy/x]\end{array} \quad \begin{array}{c}\mathcal{D}_2\\ \Gamma_2 \,\|.\ \vdash \Box\mathbf{N}t\end{array}}{\Gamma_{0,1,2} \,\|\Delta_0 \vdash A\,[t/x]}$$

where A does not contain $\Box$.

The proof of the equivalence between the rule and the axiom schema is straightforward.

Two Theorems

When a function f is defined by a system of universally closed equations E, T_f^E is the formula

$$\Box E \supset \forall x_n^1 \forall y_m^1 \left(\Box \mathbf{N} x_n^1, \mathbf{N} y_m^1 \supset \mathbf{N} f(x_n^1, y_m^1)\right)$$

where $x_n^1 \equiv x_1, \ldots, x_n$ and $y_m^1 \equiv y_1, \ldots, y_m$.

Polytime soundness *If T_f^E is derivable in feasible arithmetic then f is polytime computable.*

Polytime completeness *If f is defined in LBC, then T_f^E is derivable in feasible arithmetic*

The relationship between the sets $\Box\mathbf{N} = \{x/\Box\mathbf{N}x\}$, $\mathbf{N} = \{x/\mathbf{N}x\}$, the safe and normal positions in a function f defined in LBC and Leivant's ramified, or stratified, recursion can be summarized in the table

Normal	Safe	*LBC*
$\Box\mathbf{N}$	$\mathbf{N}$	Feasible arithmetic
$\mathbb{N}_0$	$\mathbb{N}_1$	Ramified recursion

Since there is a proof of $\Box\mathbf{N} \subseteq \mathbf{N}$ in *FA* but none of the converse inclusion, it could be argued that, from a feasible point of view, the set of feasible numbers is a subset of the natural numbers. But this supposes to identify $\Box\mathbf{N}$ with $\mathbb{N}$ and this is indeed very problematic.

A Problematic Justification of Induction

It is well known that the usual introduction and elimination rules for $\mathbf{N}$ can be justified by local reduction and expansion rules. For example, we have

$$\begin{array}{c} \vdots \\ \Gamma \vdash \mathbf{N}t \end{array} \Rightarrow_\eta \frac{\vdash \mathbf{N}0 \quad \dfrac{\mathbf{N}y \vdash \mathbf{N}y}{\mathbf{N}y \vdash \mathbf{N}Sy} \quad \begin{array}{c}\vdots \\ \Gamma \vdash \mathbf{N}t\end{array}}{\Gamma \vdash \mathbf{N}t}$$

This is impossible in the case of $\Box\mathbf{N}$ since in its elimination rule the inductive formula must be $\Box$ free. This is indeed a very important point but its discussion can be postponed until Sect. 5.3.3 since a very close situation will be encountered there.

A Problematic Feasibility

A proof theoretic justification of the rules of feasible logic at least has been given. But the problem is that the derivations in this logic are not feasible. As surprising as it may be, this is an immediate consequence of the existence of a translation of intuitionnistic logic in the former which is, as remarked above, affine linear logic $(A \Rightarrow B)* \equiv \Box A* \supset B* \equiv !A* \multimap B*$.

And of course some normal derivations in the latter can be of superexponential length.[5] Despite the fact that the proofs of the polytime soundness and completeness of feasible arithmetic are given by means of a realisability interpretation of the derivations in *FA* into the terms of a typed lambda calculus, it must be stressed that the existence of a polytime algorithm is not given by the usual computation rules operating on the terms of this calculus (β reduction). It is rather another interpretation of these typed terms in a polytime categorical model that associates to each typed term a polytime algorithm (Hofmann 2000). In this paper Hofmann introduces first a set-theoretic interpretation of the terms inductively defined.[6] As he emphasises:

> The purpose of this set-theoretic semantics is to specify the meaning of the terms. It allows us to do without any notion of term rewriting or evaluation. Of course, by directing the defining equations of the recursors one obtains a normalising rewrite system wich computes the set-theoretic meaning of first-order functions. However, there is no reason why such rewrite system should terminate in polynomial time. In order to obtain polynomial time algorithms from terms one must rather study the soundness proof [for the categorical polytime interpretation] we give and from it extract a compiler wich transforms programs [terms] of first-order type into polytime algorithms. (123)

5.3.3 *Light Affine Arithmetic*

Before we introduce a presentation of some of the rules of light affine arithmetic in the style of Davies and Pfenning (2001), something must be said about the main inspiration of light logic that could be given as a slogan: prevent diagonalization! (Baillot and Mazza 2008).

5.3.3.1 Diagonalization and Contraction

Since the first known use of a diagonal argument in du Bois-Reymond (1875), diagonalization is a tool used to produce new functions from a list of previously defined functions. Suppose that $f_m(n) = n^m$. Then for each $m \in \mathbb{N}$, f_m is polytime.

---

[5]It is a routine exercise to give, by use of the formulas $\Box A \supset A$ or $(\Box A \wedge \Box B) \supset \Box(A \wedge B)$, which are derivable in feasible logic, examples of derivations of length n with cut-free forms of exponential length n^n.

[6]For example, $\{(\lambda x : A.t)\, b\}_I \equiv \{\lambda x : A.t\}_I\, \{b\}_I \equiv \{t\}_{I[x \rightarrow \{b\}_I]}$. This is just β reduction in set-theoretic clothing, hence Hofmann's remark that rewrite rules could be used to give the meaning of the terms. The problem is that, when computed with these rules, the proof terms are not polytime algorithms.

But it is clear that the function g defined by $g(n) = f_n(n)$ is of exponential growth. From a logical, or structural, point of view, this definition involves the contraction of two hypotheses.

$$\dfrac{\begin{array}{c}\vdots\\ x:\mathbb{N}, y:\mathbb{N} \vdash x^y:\mathbb{N}\end{array}}{z:\mathbb{N} \vdash z^z:\mathbb{N}}$$

As is well known, diagonalization is also the tool that Russell borrowed from Cantor and that led him to discover his infamous proof of a contradiction in Frege's Logic. When this proof is formalised with Prawitz's rules of introduction and elimination for $\in$ given above in 1.2, the derivation obtained has exactly the structure of a non-normalisable proof term $(v)\,v, v \equiv \lambda x\,(x)\,xx$. This derivation also involves the use of the contraction rule.

The point of departure in Girard (1998) is a careful study of the axioms and rules that could allow the production of "proofs" of contradiction with Prawitz's rules. This analysis leads Girard to introduce in the same paper light linear logic. The main result given there was the proof the existence of a polytime procedure of cut elimination by use of the technology of proof nets.[7] This result was obtained by adding a modal operator § to the "of course" operator ! of linear logic and by restricting the rules of linear logic. The weakening rule was added to a slightly simplified version of this logic by Asperti and the resulting logic, affine light logic, was studied in a joint paper with Roversi where, in the spirit of the Curry Howard correspondence, a rather complicated system of proof terms was introduced (Asperti 1998; Asperti and Roversi 2002). Terui gave a much simpler system of proof terms, light affine lambda calculus, in Terui (2001) and Terui (2007). One point must be stressed. Light affine logic is intrinsically polytime: all cut elimination strategies are polytime and strongly convergent.[8]

5.3.3.2 Light Affine Logic

Our formulation of the rules of light affine logic (*LAL*) owes much to Baillot's and Terui's one in (2004; 2009). Since we want to stay as close as possible to the style of formulation used for feasible logic in 2.2, the left hand side of a sequent is separated

---

[7] We must also mention the introduction of a light set theory that has been investigated by Terui in (2004).

[8] One peculiarity of the rewriting rules of this calculus is that it enjoys a strong and convergent polytime normalisation procedure. Independently of the strategy of evaluation, the normal form of each of its terms t can be obtained in $O(|t|^{2^{d(t)+1}})$ number of steps, where $|t|$ is the length of t and $d(t)$ is the maximum of the numbers of modal operator enclosing each subterm of t in t. This result implies that the cut elimination procedure for light affine logic is also polytime and convergent independently of the strategy of cut elimination.

in three different areas. The first one on the left contains formulas that express perennial propositions that can be contracted and reused as many times as wanted; they may be think as implicitly prefixed by !. The third one on the right contains formulas that are simply true. The middle one is much more difficult to understand intuitively. For the moment, all that can be said is that the formulas in this area are implicitly prefixed with §. Those formulas are stratified and, in a sense, § can be seen as the result of a split of each of the operators !, □ of affine linear, or feasible, logic in two different operators in light affine logic !, §. Some of the modal laws that hold in affine, or feasible, logic for ! or □ , for example $!A \otimes !B \multimap !(A \otimes B)$, $!A \multimap A$ are not derivable anymore in *LAL*, but they hold for §; for example $\S A \otimes \S B \multimap \S(A \otimes B)$, $!A \multimap \S A$ are derivable.[9] Another example is the modal law $!(A \multimap B) \multimap (!A \multimap !B)$; it is not derivable in LAL, but instead there is a derivation of $\S(A \multimap B) \multimap (\S A \multimap \S B)$.

Sequent $\overbrace{G_1, \ldots, G_k}^{\text{perennial}} \parallel \overbrace{D_1, \ldots, D_m}^{\text{stratified}} \mid \overbrace{T_1, \ldots, T_n}^{\text{true}} \vdash A \text{ true}$

Axioms $. \parallel . \mid A \vdash A$

5.3.3.3 Structural Rules

Contraction is allowed only for formulas in the perennial area, weakening for all formulas.

$$\frac{\Gamma \parallel \Delta \mid \Theta \vdash C}{\Gamma, A \parallel \Delta, B \mid \Theta, D \vdash C}\,(weak) \qquad \frac{\Gamma, A, A \parallel \Delta \mid \Theta \vdash C}{\Gamma, A \parallel \Delta \mid \Theta \vdash C}\,(cont)$$

5.3.3.4 Logical Rules

The only active formulas are the true ones in the third area of the antecedent.

$$\frac{\Gamma_1 \parallel \Delta_1 \mid \Theta_1 \vdash A \quad \Gamma_2 \parallel \Delta_2 \mid \Theta_2 \vdash B}{\Gamma_{1,2} \parallel \Delta_{1,2} \mid \Theta_{1,2} \vdash A \otimes B}\,(\otimes i)$$

$$\frac{\Gamma_1 \parallel \Delta_1 \mid \Theta_1 \vdash A \otimes B \quad \Gamma_2 \parallel \Delta_2 \mid \Theta_2, A, B \vdash C}{\Gamma_{1,2} \parallel \Delta_{1,2} \mid \Theta_{1,2} \vdash C}\,(\otimes e)$$

$$\left\{\frac{\Gamma \parallel \Delta \mid \Theta \vdash A_i}{\Gamma \parallel \Delta \mid \Theta \vdash A_1 \oplus A_2}\,(\oplus i)\right\}^{1 \leqslant i < 2}$$

[9]Notice that those formulas are the ones mentioned in note 4.

$$\frac{\Gamma_{0,} \parallel \Delta_{0,} \mid \Theta_{0,} \vdash A_1 \oplus A_2 \quad \overset{1\leqslant i\leqslant 2}{\{\Gamma_i \parallel \Delta_i \mid \Theta_i, A_i \vdash C\}}}{\Gamma_{0,1,2} \parallel \Delta_{0,1,2} \mid \Theta_{0,1,2} \vdash C}\ (\oplus e)$$

$$\frac{\Gamma \parallel \Delta \mid \Theta, A \vdash B}{\Gamma \parallel \Delta \mid \Theta \vdash A \multimap B}\ (\multimap i)$$

$$\frac{\Gamma_1 \parallel \Delta_1 \mid \Theta_1 \vdash A \multimap B \quad \Gamma_2 \parallel \Delta_2 \mid \Theta_2 \vdash A}{\Gamma_{1,2} \parallel \Delta_{1,2} \mid \Theta_{1,2} \vdash B}\ (\multimap e)$$

5.3.3.5 Modal Rules

In one step of introduction for §, some, or all, of the true formulas can move to the perennial area and thus become eligible for contraction.

$$\frac{. \parallel . \mid \Gamma, \Delta \vdash A}{\Gamma \parallel \Delta \mid . \vdash \S A}\ (\S i) \qquad \frac{\Gamma_1 \parallel \Delta_1 \mid \Theta_1 \vdash \S A \quad \Gamma_2 \parallel \Delta_2, A \mid \Theta_2 \vdash C}{\Gamma_{1,2} \parallel \Delta_{1,2} \mid \Theta_{1,2} \vdash C}\ (\S e)$$

In one step of introduction for ! the formula B that moves to the perennial area, may be absent.

$$\frac{. \parallel . \mid B \vdash A}{B \parallel . \mid . \vdash !A}\ (!i) \qquad \frac{\Gamma_1 \parallel \Delta_1 \mid \Theta_1 \vdash !A \quad \Gamma_2, A \parallel \Delta_2 \mid \Theta_2 \vdash C}{\Gamma_{1,2} \parallel \Delta_{1,2} \mid \Theta_{1,2} \vdash C}\ (!e)$$

The elimination rules rests on the idea that the formulas in the modal areas are implicitly prefixed by a modal operator. Indeed, the rules

$$\downarrow \frac{\Gamma, A \parallel \Delta, B \mid \Theta \vdash C}{\Gamma \parallel \Delta \mid \Theta, !A, \S B \vdash C} \uparrow$$

are derivable.

5.3.3.6 Justification of the Rules

If a proposition A appears in the perennial (respectively stratified) area in the antecedent of a sequent, then any proof of this sequent must contain at least one step of !-introduction (respectively §-introduction) on at least one branch that links this judgment and an axiom $\parallel . \mid A \vdash A$. The justification of the introduction and elimination rules for each logical constants can be given by reduction and expansion rules that operate on proofs. For example, in the cases of the modal connectives we have

- $$\dfrac{\dfrac{\begin{array}{c}\mathcal{D}\\ .\,\|.\,|B\ \vdash A\end{array}}{B\,\|.\,|.\ \vdash !A}\quad \begin{array}{c}\overline{\|.\,|A\ \vdash A}\\ \vdots\\ \dfrac{.\,\|.\,|A\ \vdash C'}{A\,\|.\,|.\ \vdash !C'}\\ \vdots\\ \Gamma, A\,\|\Delta\,|\Theta\ \vdash C\end{array}}{\Gamma, B\,\|\Delta\,|\Theta\ \vdash C} \quad \rightarrow_\beta \quad \begin{array}{c}\mathcal{D}\\ .\,\|.\,|B\ \vdash A\\ \vdots\\ \dfrac{.\,\|.\,|B\ \vdash C'}{B\,\|.\,|.\ \vdash \S C'}\\ \vdots\\ \Gamma, B\,\|\Delta\,|\Theta\ \vdash C\end{array}$$

- $$\begin{array}{c}\vdots\\ \Gamma\,\|\Delta\,|\Theta\ \vdash !C\end{array} \quad \Rightarrow_\eta \quad \dfrac{\begin{array}{c}\vdots\\ \Gamma\,\|\Delta\,|\Theta\ \vdash !C\end{array} \quad (!i)\dfrac{\overline{.\,\|.\,|C\ \vdash C}}{C\,\|.\,|.\ \vdash !C}}{\Gamma\,\|\Delta\,|\Theta\ \vdash !C}\ (!e)$$

- $$\dfrac{\dfrac{\begin{array}{c}\mathcal{D}\\ .\,\|.\,|B\ \vdash A\end{array}}{B\,\|.\,|.\ \vdash !A}\quad \begin{array}{c}\overline{\|.\,|A\ \vdash A}\\ \vdots\\ \dfrac{.\,\|.\,|\Gamma', A, \Delta'\ \vdash C'}{\Gamma', A\,\|\Delta'\,|.\ \vdash \S C'}\\ \vdots\\ \Gamma, A\,\|\Delta\,|\Theta\ \vdash C\end{array}}{\Gamma, B\,\|\Delta\,|\Theta\ \vdash C} \quad \rightarrow_\beta \quad \begin{array}{c}\mathcal{D}\\ .\,\|.\,|B\ \vdash A\\ \vdots\\ \dfrac{.\,\|.\,|\Gamma', B, \Delta'\ \vdash C'}{\Gamma', B\,\|\Delta'\,|.\ \vdash \S C'}\\ \vdots\\ \Gamma, B\,\|\Delta\,|\Theta\ \vdash C\end{array}$$

- $$\begin{array}{c}\vdots\\ \Gamma\,\|\Delta\,|\Theta\ \vdash \S C\end{array} \quad \Rightarrow_\eta \quad \dfrac{\begin{array}{c}\vdots\\ \Gamma\,\|\Delta\,|\Theta\ \vdash \S C\end{array} \quad (\S i)\dfrac{\overline{.\,\|.\,|C\ \vdash C}}{.\,\|C\,|.\ \vdash \S C}}{\Gamma\,\|\Delta\,|\Theta\ \vdash !C}\ (\S e)$$

If we define $A \to B \equiv !A \multimap B$ as in Baillot and Terui (2004, 2009) we obtain a new connective that should not be confused with intuitionistic implication $\Rightarrow$ as shown by the derived elimination rules for $\to$.

$$\dfrac{\Gamma, A\,\|\Delta\,|\Theta\ \vdash B}{\Gamma\,\|\Delta\,|\Theta\ \vdash A \to B}\ (\to i) \qquad \dfrac{\Gamma\,\|\Delta\,|\Theta\ \vdash A \to B \quad .\,\|.\,|C\ \vdash A}{\Gamma, C\,\|\Delta\,|\Theta\ \vdash B}\ (\to e)$$

The right premiss of the elimination rule is weaker than the one for $\Rightarrow$. Nevertheless, it can be easily checked that the rules for $\to$ are justified.

5.3.3.7 Rules for N

Introduction Rules for N

$$\overline{.\,\|.\,|.\ \vdash \mathbf{N}0}\ (\mathbf{N}i) \qquad \dfrac{\Gamma\,\|\Delta\,|\Theta\ \vdash \mathbf{N}t}{\Gamma\,\|\Delta\,|\Theta\ \vdash \mathbf{N}St}\ (\mathbf{N}i)$$

Elimination Rules for N

Usually, light linear or affine logic are second order-systems and the type of light natural numbers is defined by $\mathbf{N}(x) \equiv \forall X\, [!\forall y\, (Xy \multimap XSy) \multimap \S\, (X0 \multimap Xx)]$ or, equivalently, by $\mathbf{N}(x) \equiv \forall X\, [!\forall y\, (Xy \multimap XSy) \multimap (\S X0 \multimap \S Xx)]$. Since $\mathbf{N}(x) \vdash !\forall y\, (A[y/x] \multimap A[Sy/x]) \multimap (\S A[0/x] \multimap \S A[y/x])$, it is straightforward to extract from this second-order definition a weaker first-order induction principle formulated as an elimination rule for **N**.

$$\dfrac{\|.\mid \Gamma_0, \Delta_0 \vdash A[0/x] \qquad \begin{array}{c} .\,\|.\,| A[y/x] \vdash A[y/x] \\ \mathcal{D} \\ .\,\|.\,| B, A[y/x] \vdash A[Sy/x] \end{array} \qquad \Gamma_1 \| \Delta_1 \mid \Theta_1 \vdash \mathbf{N}t}{B, \Gamma_{0,1} \| \Delta_{0,1} \mid \Theta_1 \vdash \S A[t/x]}\ (\mathbf{N}e)$$

Justification of the Rules for N

The β reduction rules are:

$$\dfrac{.\,\|.\mid \Gamma_0, \Delta_0 \vdash A[0/x] \qquad \begin{array}{c} .\,\|.\,| A[y/x] \vdash A[y/x] \\ \mathcal{D} \\ .\,\|.\,| B, A[y/x] \vdash A[Sy/x] \end{array} \qquad \begin{array}{c} \vdots \\ \Gamma_1 \| \Delta_1 \mid \Theta_1 \vdash \mathbf{N}0 \end{array}}{B, \Gamma_{0,1} \| \Delta_{0,1} \mid \Theta_1 \vdash \S A[0/x]}$$

$$\to_\beta$$

$$\dfrac{\dfrac{\begin{array}{c} \vdots \\ .\,\|.\mid \Gamma_0, \Delta_0 \vdash A[0/x] \end{array}}{\Gamma_0 \| \Delta_0 \mid . \vdash \S A[0/x]}\ (\S i)}{B, \Gamma_{0,1} \| \Delta_{0,1} \mid \Theta_1 \vdash \S A[0/x]}\ (weak)$$

$$\dfrac{\begin{array}{c} \vdots \\ \|.\mid \Gamma_0, \Delta_0 \vdash A[0/x] \end{array} \qquad \begin{array}{c} .\,\|.\,| A[y/x] \vdash A[y/x] \\ \mathcal{D} \\ .\,\|.\,| B, A[y/x] \vdash A[Sy/x] \end{array} \qquad \dfrac{\begin{array}{c} \vdots \\ \Gamma_1 \| \Delta_1 \mid \Theta_1 \vdash \mathbf{N}t \end{array}}{\Gamma_1 \| \Delta_1 \mid \Theta_1 \vdash \mathbf{N}St}}{B, \Gamma_{0,1} \| \Delta_{0,1} \mid \Theta_1 \vdash \S A[St/x]}$$

$$\to_\beta$$

$$\dfrac{\dfrac{\begin{array}{c} \vdots \\ B, \Gamma_{0,1} \| \Delta_{0,1} \mid \Theta_1 \vdash \S A[t/x] \end{array} \qquad (\S i)\ \dfrac{\begin{array}{c} .\,\|.\,| A[t/x] \vdash A[t/x] \\ \mathcal{D}[t/x] \\ .\,\|.\,| B, A[t/x] \vdash A[St/x] \end{array}}{B \| A[t/x] \,|. \vdash \S A[St/x]}}{B, B, \Gamma_{0,1} \| \Delta_{0,1} \mid \Theta_1 \vdash \S A[St/x]}\ (\S e)}{B, \Gamma_{0,1} \| \Delta_{0,1} \mid \Theta_1 \vdash \S A[St/x]}\ (cont)$$

But, as in the case of feasible induction, there is no expansion rule for **N**, since one elimination step for this constant involves a stratification of the induction formula. Indeed this shows that the introduction rules are stronger than the elimination rules. This should not be a big surprise. Usually, the introduction rules allow to build the normal form of any natural number $S \dots S0$, and the elimination rules to prove $\mathbf{N}y \rightarrow \mathbf{N}t\,[y/x]$, for a term t. Obviously, for some term t such proofs must be prohibited if the system is polytime. To put it differently, all numbers as values are available in light affine arithmetic but for numbers defined as the value of a potential computation $t\,[n/x]$ it must be impossible to prove $\mathbf{N}t\,[n/x]$ if the computation is not feasible.

The most serious problem is the connective §that can be justified, at least at first sight, only a pragmatic ground: the rules for §give very useful tools to control the computational complexity of the cut elimination procedure. But, despite the fact that the rules that define this operator are justified in the sense of proof-theoretic semantics, some more convincing arguments should be given that could explain the meaning of this operator. Let us begin with a modest one that may help to understand it in the case of the constant **N**. We must notice that we have a proof of a coercion principle $\forall x\,(\mathbf{N}\,(x) \multimap \S\,\dagger^k\,\mathbf{N}\,(x))$ for $\dagger \in \{!, \S\}$ but no proof of $\forall x\,(\S^k\mathbf{N}x \multimap \mathbf{N}x)$, $k \geq 1$. This remind us of the distinction between $\square$**N** and **N** in feasible arithmetic that reflects the one between normal and safe variables in Bellantoni and Cook definition. This very informal analogy has been made much more precise in Murawski and Ong (2004) where a subsystem LBC^- of Bellantoni's, Cook's and Leivant's LBC is introduced. Then, a compositional translation LBC^- into second order light affine logic is defined such that the following theorem holds.

Theorem. *If f of type $\overbrace{\mathbb{N}_0, \dots, \mathbb{N}_0}^{n}, \overbrace{\mathbb{N}_1, \dots, \mathbb{N}_1}^{m} \rightarrow \mathbb{N}_1$ is defined in LBC^-, then the compositional translation of the definition of f is a proof in light affine logic of the sequent*

$$\mathbf{N}\,(x_1)\,, \dots, \mathbf{N}\,(x_n)\,, \S^k\mathbf{N}\,(y_1)\,, \dots, \S^k\mathbf{N}\,(y_m) \vdash \S^k\mathbf{N}f\,\left(x_n^1, y_m^1\right).$$

Thus the table given in Sect. 5.3.2 above can now be completed

$\square$**N**	**N**	Feasible arithmetic
$\mathbb{N}_0$ Normal	$\mathbb{N}_1$ Safe	***LBC***
N	$\S^k$**N**	Light affine logic

5.4 Conclusion

A proof-theoretic semantics for feasible logic and first-order light affine logic has been outlined, even if the first one is not really a feasible logic. In both cases, the justification of the induction principle by an η expansion rule fails. This shows that the main restriction that must be made to constructive arithmetic in order to obtain a feasible arithmetic is the adoption of a weak induction principle. In the case of light arithmetic this involves the use of a modal connective that creates for each proposition its stratified counterpart that can be seen as a way to control, if not to avoid completely, the undesirable consequences of impredicative definitions. In future work it could be interesting to study with an eye on proof-theoretic semantics light set theory where a highly impredicative comprehension principle is used (Girard 1998; Terui 2004). Since it is well known that feasible and light systems are intensionally very weak even if they are polytime extensionally complete, the non-size increasing system that allows definition of polytime algorithms that are forbidden in the former deserve also to be taken in account (Aehlig et al. 2004; Hofmann 2003).

References

Aehlig, K., Berger, U., Hofmann, M., & Schwichtenberg, H. (2004). An arithmetic for non-size-increasing polynomial-time computation. *Theoretical Computer Science, 318*, (1–2), 3–27.

Asperti, A. (1998). Light affine logic. In *Proceedings of LICS'98*, Indianapolis (pp. 300–308). IEEE Computer Society.

Asperti, A., & Roversi, L. (2002). Intuitionnistic light affine logic. *ACM Transactions on Computational Logic, 3*(1), 1–39.

Baillot, P., & Mazza, D. (2010). Linear logic by levels and bounded time complexity. *Theoretical Computer Science, 411*(2), 470–503.

Baillot, P., & Terui, K. (2004). Light types for polynomial time computation in lambda calculus. In *Proceedings of LICS'04*, Turku (pp. 266–275).

Baillot, G., & Terui, K. (2009). Light types for polynomial time computation in lambda calculus (long version). *Information and Computation, 207*(1), 41–62.

Bellantoni, S., & Cook, S. (1997). New recursion-theoretic characterization of the polytime functions. *Computationnal Complexity, 2*, 97–110.

Bellantoni, S., & Hofmann, M. (2002). A new "feasible" arithmetic. *The Journal of Symbolic Logic, 67*, 104–116.

du Bois-Reymond, P. (1875). Ueber asymptotische Werke, infinitäre Approximationen und infinitäre Auflösung von Gleichungen. *Mathematische Annalen, 8*, 363–414.

Brouwer, L. E. J. (1948). Essentieel negatieve eigenschappen. *KNAW, 51*, 963–964. *Indagationes Mathematicae, 10*, 322–323. [A. Heyting (Ed., Trans.) *Collected Works* 1, 478–479]

Borel, E. (1950). *Leçons sur la théorie des fonctions*, Gauthier-Villars, 4ème édition, 1ère édition 1898.

Bishop, E. (1967). *Foundations of constructive analysis*. New York: McGraw-Hill.

Buss, S. (1987). On Gödel's theorem on length of proofs I: Number of lines and speed up for arithmetics. *The Journal of Symbolic Logic, 52*(4), 916–927.

Cobham, A. (1965). The intrinsic computational difficulty of functions. In Y. Bar-Hellel (Ed.), *Proceedings of the 1964 international congress for logic, methodology, and philosophy of sciences* (pp. 24–30). Amsterdam: North Holland.

Curry, H. B., & Feys, R. (1958). *Combinatory logic* (Vol. I). Amsterdam: North-Holland.

van Dantzig, D. (1956). Is $10^{10^{10}}$ a finite number? *Dialectica, 9*, 273–277.

Davies, R., & Pfenning, F. (2001). A judgmental reconstruction of modal logic. *Mathematical Structures in Computer Science, 11*(4), 511–540.

Dubucs, J. (1997). Logique, effectivité et faisabilité. *Dialogue, XXXVI*, 45–68.

Dubucs, J., & Marion, M. (2003). Radical anti-realism and substructural logics. In A. Rojszczak, J. Cachro, & G. Kurczewski (Eds.), *Philosophical dimensions on logic and science, selected contributed papers from the 11th international congress of logic, methodology and philosophy of science*, Krakow (pp. 235–249). Kluwer.
Dummett, M. A. E. (1975). The philosophical basis of intuitionnistic logic. In H. E. Rose, & J. Schepherdson (Eds.), *Logic colloquium* (Vol. 73, pp. 5–40). Amsterdam: North-Holland.
Gentzen, G. (1935). Untersuchungen über das logiesche Schliessen. *Mathematishe Zeitschrift, 39*, 176–210, 405–431. [M. E. Szabo (Ed., Trans.) *The collected papers of Gerhard Gentzen* (Studies in logic and the foundations of mathematics, pp. 68–131). Amsterdam: North-Holland]
Girard, J. Y. (1998). Light linear logic. *Information and Computation, 14*(3), 175–204.
Gödel, K. (1936). Über die Länge von Beweisen. Ergebnisse eines Mathematischen Kolloquiums 7, (pp. 23–24); Feferman, S. et al. (Eds.), *Kurt Gödel Collected Works* (vol. 1, pp. 306–309). Oxford: Oxford University Press.
Gödel, K. (1958). Über ein bisher noch nicht benütze Erweiterung des finiten Standpunktes. *Dialectica, 12*, 280–287; Feferman, S. et al. (Eds.), *Kurt Gödel Collected Works* (vol. 2, pp. 240–251). Oxford: Oxford University Press.
Gödel, K. (1990). On an extension of finitary mathematics which has not yet been used. In S. Feferman et al. (Eds.), *Kurt Gödel Collected Works* (Vol. 2., pp. 271–280). Oxford: Oxford University Press.
Hofmann, M. (2000). Safe recursion with higher types and BCK-algebra. *Annals of Pure and Applied Logic, 104*, 1–3, 113–166.
Hofmann, M. (2003). Linear types and non-size-increasing polynomial time computation. *Information and Computation, 183*(1), 57–85.
Howard, W. A. (1980). The formulae-as-types notion of construction. In J. P. Seldin & J. R. Hindley (Eds.), H.B. Curry: *Essays on combinatory logic, lambda-calculus and formalism* (pp. 479–480). New York: Accademic Press.
Kleene, S. C. (1973). Realisability: A retrospective survey. In H. Rodgers & A. R. D. Mathias (Eds.), *Cambridge summer school in mathematical logic 1971* (Lecture notes in mathematics, vol. 337, pp. 95–112). Berlin/New York: Springer.
Leivant, D. (1993). Stratified functional programs and computational complexity. In *Proceedings of the 20th symposium on principles of programming languages*, Charleston (pp. 325–333). ACM.
Martin-Löf, P. (1984). *Intuitionistic type theory, notes by Giovanni Sambin on a serie of lectures given in Padua*, June 1980. Bibliopolis: Napoli.
Martin-Löf, P. (1985). On the meaning of the logical constants and the justification of the logical laws [Attidegli incontri di logical mathematica, 1983, Vol. 2. Dipartimento di Mathematica, Universita di Siena].
Martin-Löf, P. (1987). Truth of a proposition, evidence of a judgement, validity of a proof. *Synthèse, 73*, 407–420.
Murawski, A. S., & Ong, C.-H. L. (2004). On an interpretation of safe recursion in light affine logic. *Theoretical Computer Science, 318*(1–2), 197–223.
Prawitz, D. (1965). *Natural deduction, a proof-theoretical study* (Stockolm studies in philosophy, vol. 3). Stockholm: Almqvist & Wiksell.
Shapiro, S. (1985). Epistemic and intuitionistic arithmetic. In S. Shapiro (Ed.), *Intensional mathematics* (Studies in logic and the foundations of mathematics, vol. 113, pp. 11–46). Amsterdam: North-Holland.
Terui, K. (2001). Light affine lambda calculus and polytime strong normalisation. In *Proceedings of LICS'01*, Boston (pp. 209–220). IEEE Computer Society.
Terui, K. (2004). Light affine set theory; A naive set theory of polynomial time. *Studia Logica, 77*, 9–40.
Terui, K. (2007). Light affine lambda calculus and polynomial time strong normalization. *Archive for Mathematical Logic, 46*(3–4), 253–280.

Troelstra, A. S., & van Dalen, D. (1988). *Constructivism in mathematics. An introduction, I&II* (2 vols.). Amsterdam: North-Holland.

Turing, A. (1936–37). *On computable numbers, with an application to the entscheidungsproblem* (Proceedings of the london mathematical society, ser. 2, Vol. 42, pp. 230–265). Corrections 1937, ibid., vol. 43, pp. 544–546.

Wright, C. (1982). Strict finitism. *Synthese, 51*(2), 203–282.

Chapter 6
Recursive Functions and Constructive Mathematics

Thierry Coquand

6.1 Introduction

The goal of this paper is to discuss the following question: is the theory of recursive functions needed for a rigorous development of constructive mathematics? I will try to present the point of view of constructive mathematics on this question. The plan is the following: I first explain the gradual loss of appreciation of constructivity after 1936, clearly observed by Heyting and Skolem, in connection with the development of recursivity. There is an important change in 1967, publication of Bishop's book, and the (re)discovery that the theory of recursive functions is actually *not* needed for a rigorous development of constructive mathematics. I then end with a presentation of the current view of constructive mathematics: mathematics done using intuitionistic logic, view which, surprisingly, does not rely on any explicit notion of algorithm.

6.2 Recursive Function Theory

Let us first recall the definition of recursive function, as done in classical mathematics. A numerical function is *recursive* if it can be defined by the usual schemas defining *primitive recursive functions* (projections, constant functions, composition, primitive recursion) and the following clause: if $g(u_1, \ldots, u_n, x)$ is a recursive functions such that for all $u_1, \ldots, u_n$ there exists x such that

T. Coquand (✉)
Chalmers Tekniska Högskola och, Göteborgs Universitet, 412 96 Göteborg, Suède
e-mail: coquand@chalmers.se

J. Dubucs, M. Bourdeau (eds.), *Constructivity and Computability in Historical and Philosophical Perspective*, Logic, Epistemology, and the Unity of Science 34,
DOI 10.1007/978-94-017-9217-2_6

$g(u_1, \ldots, u_n, x) = 0$ then the function f which to $u_1, \ldots, u_n$ associates the least x such that $g(u_1, \ldots, u_n, x) = 0$ is a recursive function.

Classically, this definition is seen as the exact mathematical definition of the intuitive notion of computable functions.[1] In this paper, we want to discuss the opinion that this notion is *necessary* for a precise discussion of the notion of functions in constructive mathematics. This opinion is well expressed is the following citation from Sanin (1958) that states that constructive mathematics "began to be developed successfully only in the middle of the 1930s after the precise mathematical concept of *arithmetic algorithm (computable arithmetic function)* had been worked out. Only the introduction into mathematics of the precise notion of arithmetic algorithm created a satisfactory basis for the treatment of the constructive interpretation of mathematical propositions and fundamental notions of constructive mathematical analysis." Thus the opinion is that, in order to develop in a rigorous way constructive mathematics, one needs to have a mathematically precise notion of computable function, and that this is provided by the notion of recursive functions.

There are several problems with such claim, that have been clearly discussed by Heyting and Skolem (and that we shall recall here, since these are seldom discussed). These problems are connected with the meaning of existence in the last clause of definitions of recursive functions: "for all $u_1, \ldots, u_n$ there *exists* x such that $g(u_1, \ldots, u_n, x) = 0$".[2]

6.2.1 Heyting

Heyting (1962) addresses exactly this question of the meaning of the existential quantifier in the definition of recursive functions. "As soon as the existential quantifier in the definition is interpreted non constructively, the notion of recursive function is no longer connected with that of a calculable function. On the other hand, if the quantifier is interpreted constructively, then the definition presupposes some notion of a calculable function." This is a serious objection to the conceptual importance of the notion of recursive function in constructive mathematics. It shows that, understood constructively, the clauses for recursive functions cannot be used to *define* the notion of computable function, since this notion is required to understand the meaning of a sentence of the form $\forall x \exists y\ R(x, y)$.

Heyting (1962) also complains about the problem of not being precise in the metalogic (intuitionistic or classical) used with this notion of recursive function. "The good habit of distinguishing between results on recursive functions obtained by intuitionistic logic and those which for their proof need classical logic is

---

[1]As is well known, there are several different equivalent form of this definition, but the discussion that follows apply for these variants as well.

[2]We stress that we limit our discussion to the notion of *total* recursive functions, and do not consider *partial* recursive functions, introduced by Kleene. Only total recursive functions are relevant for an explanation of the notion of functions in constructive mathematics, which are total by definition Bishop (1967) and Richman (1990).

abandoned in many recent papers and books. I regret this, because thereby the connection of the theory with the notion of effective calculability is obscured."

The problem in the use of classical logic at the metalevel with this definition of recursive function is clearly shown by the following example. It is direct that all functions $g_k(n) = n < k$ for $k = 0, 1, \ldots$ and the constant function $g(n) = 1$ are recursive. If we define $f(n) = 1$ iff there is n consecutive 7 in the decimal development of π and $f(n) = 0$ otherwise, then, classically, f is either one of these functions g_k or is the constant function g. In both cases, f is a recursive function. Hence, classically, f is a recursive function. But this does not seem satisfactory since, intuitively, we have no way to compute $f(n)$ for a general n.[3]

As noted by Heyting (1958), Kleene himself (1952) remarks that "we should not claim that a function is effectively calculable on the ground that it has been shown to be general recursive, unless the demonstration that it is general recursive is effective", and Heyting adds that "this means that for the latter demonstration an intuitive notion of effectiveness is indispensable". He also points out in a short paper (Heyting 1957) the relevant comment of Rosza Peter that "any attempt to define the notion of a constructive theory leads into a vicious circle, because the definition always contains an existential quantifier, which in its turn must be interpreted constructively".

6.2.2 Skolem

Skolem has a similar concern about the previous definition of recursive function (Skolem 1955). "At any rate one should not use the quantifiers without any closer explanation of how they are to be interpreted ... If we for instance consider an arithmetical sentence of the form $\forall x.\exists y.R(x, y)$ we may interpret it as follows: We know for any x how to find a y such that $R(x, y)$. I had once some difficulty in reading the important paper of S.C. Kleene (1936) ... because he, as other authors, introduces the quantifiers without comments. In his article we have for example the theorem: Let $R(x, y)$ be a recursive relation and let $\forall x.\exists y.R(x, y)$ be true. Then $\mu y.R(x, y)$ is a recursive function of x. Certainly one ought to know here what is meant by the words: let $\forall x.\exists y.R(x, y)$ be true."

The concern is also in the meaning of the existence statement: for all x there *exists* y such that $R(x, y)$. It cannot be understood in a formal way, that the statement $\forall x \exists y.\ R(x, y)$ is provable in a given formal system, like ZF or ZFC or ZF extended by some large cardinal axiom, since the meaning will then depend on the formal system. How shall we understand this statement then? "Perhaps it is conceived in the Platonist way which means it is take for granted that every proposition has a meaning per se, and it is decided per se, whether it is "true" or not. I find this view unacceptable from a finitist standpoint."

---

[3]This problem does not appear using a constructive meta-level, since, in this case, we can only consider f to be a function when we can decide, for each n, whether or not there is n consecutive 7 in the decimal development of π.

In conclusion, the definition of recursive functions, which seems to capture in a precise mathematical way the intuitive notion of computable function, appears actually to lack conceptual precision. It is even simply irrelevant, when analysed from a constructive point of view.

6.2.3 *Intuitionistic Consistency Proofs: Two Examples*

We shall now illustrate the "belief" that recursive function theory is a prerequisite for having a precise notion of intuitionistically given sequence of objects in the logical literature. This is expressed in two reviews of papers that establish consistency proofs using an intuitionistic metalanguage. One is Church's review of one of a paper of P.S. Novikoff's (1943) and the other is H. Wang's review of Lorenzen's paper on the consistency of the ramified theory of types (Lorenzen 1951).

6.2.3.1 Novikoff

Novikoff's consistency proof relies on the introduction of infinitary formulae. One introduces then an inductive definition of *regularity* (similar to cut-free provability) and one shows that any provable formula is regular. This notion of infinitary formulae is an example of "generalised" inductive definitions. Besides the usual inductive clauses: if F and G are formulae, so are $\bar{F},\ FG,\ F \vee G$ and $F \rightarrow G$, there is also an infinitary one which can be stated as: if $F_1,\ F_2,\ F_3,\ \ldots$ are formulae then so are $F_1 F_2 F_3 \ldots$ and $F_1 \vee F_2 \vee F_3 \vee \ldots$ This is an early presentation of infinitary propositional calculus $L_{\omega_1,\omega}$. This inductive definition makes sense constructively (Bishop uses a similar definition for representing Borel sets Bishop 1967). The proof is correct intuitionistically and very interesting (see for instance Mints' survey article Mints 1991). However, Novikoff's paper can only be understood if one accepts to take this generalised inductive notion of formulae as primitive and constructively meaningful.

Church (1946) gave a clear and precise review of this paper. We shall be interested in the informal evaluation of the paper by Church: "In the reviewer's opinion the most serious objection to the author's formulation is the failure to specify more precisely the restriction imposed, that an infinite sequence of formulas used in the construction of a formula ... must be intuitionistically given ... This observation moreover casts doubt on the author's claim that his consistency proof is intuitionistically valid."

At the end of his paper, Novikoff presents an elegant application: a proof of closure under Markov's rule. If each $F(i)$ is either $\top$ or $\bot$ and $F(1) \vee F(2) \vee \ldots$ is provable then we can find k such that $F(k)$ is $\top$. There also, this argument can only be understood constructively. Church comments on this that "the author enters into a discussion which apparently is intended to show some bearing of his notion of regularity upon questions of existence and effective existence in recursive mathematics. But the results actually obtained in this discussion, if the reviewer has understood them correctly, are trivial or nearly so."

These comments are a clear indication of the lack of appreciation of constructivity, which was noticed in the paper of Heyting above (1962). One also can notice that another review by McKinsey of the same paper qualifies the calculus as "non constructive", probably because of its infinitary character. It is tempting to associate this lack of appreciation of constructivity to the belief that only the (classical) notion of recursive functions can give a precise meaning to the notion of constructive objects. For instance, in this view, when introducing infinitary formulae, one would talk about codes (as natural numbers) of formulae, and only consider a formula $F_1F_2F_3\ldots$ if the sequence $F_1,\ F_2,\ F_3,\ \ldots$ is recursive.

6.2.3.2 Lorenzen's Paper

Similar remarks can be made about Wang's review (1951) of Lorenzen's proof of consistency of ramified type theory (Lorenzen 1951). Wang fails to acknowledge explicitely the difference between Lorenzen's argument, which is constructive, and another consistency proof which would be by a truth definition (non constructive): "the reviewer finds it very hard to articulate about the difference between the two consistency proofs of R." As he then writes, Lorenzen "himself phrases the difference by saying that his proof satisfies Hilbert's requirement of consistency proofs that only constructive modes of reasoning are available." What is surprising is that Wang does not state explicitely that Lorenzen's proof is done in a constructive metalanguage. There is an important difference with a proof of consistency via a truth definition which is not constructively valid (since the truth of an universally quantified statement is in general not decidable). This difference is furthermore pointed out by Lorenzen, but Wang finds this distinction between a constructive argument and a non constructive one "hard to articulate".

6.3 Foundations of Constructive Analysis

The situation in 1967 (before the publication of Bishop's book (1967)) seems to be the following. The complaints of Heyting and Skolem about the lack of concern for constructive issues, are forgotten. Recursive function theory is usually presented without any consideration if the results are obtained using intuitionistic logic or not (see for instance Shoenfield 1967). Even when presented intuitionistically, the theory of recursive functions is furthermore considered to be a *necessary prerequisite* for developing constructive mathematics. Heyting, in the 1962 paper (Heyting 1962), which is a reflection on 30 years of work in the foundation of mathematics states that "constructivity is much less appreciated now than it was 30 years ago. Yet effectiveness is a primitive notion, for a proof is only a proof if it is effectively given".

A good example is provided by the book *Notes on Constructive Mathematics* by P. Martin-Löf (1970), which was, as stated in the preface, written before the author got to know about Bishop's book. This book clearly adopts the view that a rigorous presentation of constructive mathematics should rely on the notion of recursive functions. The book is also remarkable in giving a point-free presentation of some spaces, such as Cantor space. But, typically, an open set of Cantor space is defined to be a *recursively enumerable set* of basic open (that are concrete objects), and not, as we shall do it now in constructive mathematics, as an arbitrary set of basic open (leaving this notion of "arbitrary set" informal). Similarly, it presents informally the *Borel* subsets of Cantor space as infinary propositional formulae on simple sets (basic open) but feels the need to make precise this definition and define then Borel subsets of Cantor space as a recursively enumerable set of suitable finite sequences.[4] Martin-Löf remarks also that "long before the notion of effectiveness was handled with great precision by Borel (1908) and Brouwer (1918) who based many non trivial mathematical arguments on it". Most of the results in this book can be directly interpreted as important contribution to constructive mathematics without relying on the notion of recursive function. Indeed, this work is an important motivation for the introduction of formal topology (Sambin 1987). Logically, it may seem unsatisfactory to leave the notion of functions informal as done in Bishop (1967) and Mines et al. (1988). Type theory (Martin-Löf 1984) provides a precise formal system in which one can express constructive mathematics (Rathjen 2005).

6.3.1 Bishop

A real conceptual breakthrough was achieved in the work of Bishop *Foundations of Constructive Analysis* (Bishop 1967). Bishop (re)discovered that it is not only possible, but also conceptually more satisfactory, to introduce functions in constructive mathematics without mentioning recursivity. He uses the notion of "rules" as primitive in his definition of function The reasons are clearly stated (Bishop 1967). "This requirement that every sequence of integers must be recursive is wrong on three fundamental grounds. First, there is no doubt that the naive concept is basic, and the recursive concept derives whatever importance it has from some presumption that every algorithm will turn out to be recursive. Second, the mathematics is complicated rather than simplified by the restriction to recursive sequences ... Third, no gain in precision is actually gained ... the notion of a recursive function is at least as imprecise as the notion of a correct proof." This conceptual breakthrough is clearly emphasized in Stolzenberg's review of Bishop's book (Stolzenberg 1970). "The undefined concept of construction actually

[4]As noticed in Heyting's paper (1961), and as is also stressed in Kreisel's review of this paper, such infinitary notions are naturally and elegantly expressed using generalised inductive definitions in an intuitionistic metatheory.

admits a usage no less precise and clear than the undefined concept of an integer. This remarkable fact can only be obscured by bringing in recursive functions at this level."

6.3.2 *What is Then Constructive Mathematics?*

These insights of Bishop, that Turing machines and recursive functions are not relevant to explain the notion of function in constructive mathematics, are now taken into account in current work in constructive mathematics (Lombardi, Bridges, Richman, ...). For instance, Richman writes that "equivalence of Turing machines, recursive functions, and Markov algorithms presumably settled once and for all the precise definition of "computable." Bishop claimed, to the contrary, that if you had a constructive proof that a Turing machine halted and produced the desired result, then you didn't need the Turing machine in the first place; and in the absence of such a proof, the Turing machine didn't establish computability." This is reminiscent of the concerns that were made previously by Heyting and Skolem.

An interesting further conceptual discovery in current constructive mathematics has been the gradual realisation that the best description of constructive mathematics is *mathematics developed using intuitionistic logic* (Richman 1990; Bridges 1999). What is remarkable in this characterisation is that it is independent of any notion of algorithms. Another connected remarkable fact, discovered by and Heyting (1930), is that intuitionnistic logic can be captured by a formal system of deduction rules.

6.3.3 *Example*

Let us give a typical example of a statement of constructive mathematics, here constructive algebra (Mines et al. 1988; Coquand and Lombardi 2006). One defines a *commutative ring* as usual. A *local ring* will be defined as a ring such that, for any element x of the ring, we have an inverse of either x or of $1 - x$. Given these definitions, we can now state the following result.

Theorem. *If F is an idempotent square matrix over a local ring R then F is similar to a matrix of the canonical form*

$$\begin{pmatrix} I & 0 \\ 0 & 0 \end{pmatrix}$$

All this can be read as an ordinary mathematical statement. If we read this statement constructively, we can see it as the specification of an algorithm. This algorithm, given a "subprogram" witnessing the fact that given x, we can produce an inverse of x or of $1 - x$, will compute for any projection matrix F an invertible matrix P such that the projection matrix PFP^{-1} is in canonical form. The constructive proof is both the program and its correctness proof. Notice also that this "subprogram" providing an inverse of x or of $1-x$ does not need to be computable: it can be an "oracle" which produces the required inverse in a non computable way. Also, the ring operation are implicitly "computable" (since they are constructive functions), but the ring equality is not assumed to be decidable.

For the proof of this statement to be constructive (see for instance Coquand and Lombardi 2006), we do not need to mention explicitely the notion of "programs" or "algorithms". All we require is that this argument is intuitionistically valid, i.e. does not rely on the law of Excluded-Middle.

6.4 Conclusion

The world of constructive mathematics is "incomparable" conceptually to the world of recursive function theory developed in classical mathematics. One can however establish technical connections. For instance, in a fixed formal system, one can prove that any well-defined constructive function is a computable function, for instance using Kleene's realisability interpretation (Kleene 1952). But it is crucial to keep the distinction between the two notions. As explained clearly by the citations of Heyting, Skolem and Bishop above, the notion of *function* in constructive mathematics is a primitive notion which cannot be explained in a satisfactory way in term of recursivity.

Acknowledgements I would like to thank Göran Sundholm for interesting discussions on the topic of this paper and for sending me the references (Heyting 1957).

References

Bishop, E. (1967). *Foundations of constructive mathematics* (xiii+370pp). New York/Toronto/Ont.-London: McGraw-Hill.

Borel, E. (1908). Les "Paradoxes" de la théorie des ensembles. *Ann. Sci. Ecole Norm Sup, 25*, 443–338.

Bridges, D.S. (1999). Constructive mathematics: a foundation for computable analysis. *Theoretical Computer Science, 219*, 95–109.

Brouwer, L. (1918). Begründung der Mengenlehre unabhängig vom logish'chen Satz vom ausgeschlossenen Dritten. *Ester Teil. Verh. Nederl. Akad. Wetensch. Afd. Natuurk., Sct. I, 12*(5), 3–43.

Coquand, Th., & Lombardi, H. (2006). A logical approach to abstract algebra. *Mathematical Structures in Computer Science, 16*(5), 885–900.

Church, A. (1946). Review of Novikoff's (1943). *Journal of Symbolic Logic, 11*(4), 129–131.

Van Heijenoort, J. (1977). *From Frege to Gödel: a source book in mathematical logic, 1879–1931* (Vol. 9). Cambridge: Harvard University Press.

Heyting, A. (1930). Die formalen Regeln der intuitionistischen Logik. I, II, III. *Sitzungsberichte Akad. Berlin* (pp. 42–56, 57–71, 158–169).

Heyting, A. (1957). Some remarks on intuitionism. In *Proceedings of colloquium on constructivity in mathematics*, Amsterdam, pp. 69–71.

Heyting, A. (1958). Intuitionism in mathematics. In R. Klibansky (Ed.), *Philosophy in the mid-century: A survey* (Vol. 1, pp. 101–115). *Logic and philosophy of science*. Florenc: La Nuova Italia Editrice.

Heyting, A. (1961). Infinitistic methods from a finitist point of view. *Proceedings of the symposium on foundations of mathematics on infinitistic methods, Warsaw, 1959* (pp. 185–192). Oxford/London/New York/Paris: Pergamon.

Heyting, A. (1962). After thirty years. In *Logic, methodology and philosophy of science* (pp. 194–197). Stanford: Stanford University Press.

Heyting, A. (1962). After thirty years. In N. Ernest, S. Patrick, T. Alfred (Eds.), *Proceedings of the 1960 international congress on logic, methodology and philosophy of science* (pp. 194–197). Stanford: Stanford University Press.

Kleene, S. C. (1936). General recursive functions on natural numbers. *Mathematische Annalen, 112*, 727–742.

Kleene, S. C. (1952). *Introduction to metamathematics* (x+550pp). New York: D. Van Nostrand.

Lorenzen, P. (1951). Algebraische und logistische Untersuchungen über freie Verbände. *Journal of Symbolic Logic, 16*, 81–106.

Martin-Löf, P. (1970). *Notes on constructive mathematics* (109pp). Stockholm: Almqvist and Wiksell.

Martin-Löf, P. (1984). *Intuitionistic type theory*. Bibliopolis: Napoli.

Mines, R., Richman, W., & Ruitenburg, W. (1988). *A course in constructive algebra.* Heidelberg: Springer.

Mints, G. (1991). Proof theory in the USSR: 1925–1969. *Journal of Symbolic Logic, 56*(2), 385–424.

Novikoff, P. S. (1943). On the consistency of certain logical calculus. *Rec. Math. [Mat. Sbornik] N.S., 12*(54), 231–261.

Rathjen, M. (2005). The constructive Hilbert program and the limits of Martin-Löf type theory. *Synthese, 147*(1), 81–120.

Richman, F. (1990). Intuitionism as a generalisation. *Philosophia Mathematica (2), 5*, 124–128.

Sambin, G. (1987). Intuitionistic formal spaces—a first communication. *Mathematical logic and its applications (Druzhba, 1986)* (pp. 187–204). New York: Plenum.

Sanin. (1958). A constructive interpretation of mathematical judgments (Russian). *Trudy Mat. Inst. Steklov., 52*, 226–311.

Shoenfield, J. R. (1967). *Mathematical logic.* Reading: Addison-Wesley.

Skolem, T. (1955). A critical remark on foundational research. *Norske Vid. Selsk. Forh., Trondheim, 28*, 100–105.

Stolzenberg, G. (1970). Review: Foundations of constructive analysis by Errett Bishop. *Bulletin of the American Mathematical Society, 76*, 301–323.

Wang, H. (1951). Review of P. Lorenzen Algebraische und logistische Untersuchungen über freie Verbände. *Journal of Symbolic Logic, 16*, 269–272.

Chapter 7
Gödel and Intuitionism

Mark van Atten

Abstract After a brief survey of Gödel's personal contacts with Brouwer and Heyting, examples are discussed where intuitionistic ideas had a direct influence on Gödel's technical work. Then it is argued that the closest rapprochement of Gödel to intuitionism is seen in the development of the Dialectica Interpretation, during which he came to accept the notion of computable functional of finite type as primitive. It is shown that Gödel already thought of that possibility in the Princeton lectures on intuitionism of Spring 1941, and evidence is presented that he adopted it in the same year or the next, long before the publication of 1958. Draft material for the revision of the Dialectica paper is discussed in which Gödel describes the Dialectica Interpretation as being based on a new intuitionistic insight obtained by applying phenomenology, and also notes that relate the new notion of reductive proof to phenomenology. In an appendix, attention is drawn to notes from the archive according to which Gödel anticipated autonomous transfinite progressions when writing his incompleteness paper.

The principal topics are (1) personal contacts Gödel had with Brouwer and Heyting; (2) various influences of intuitionism on Gödel's work, in particular on the introduction of computable functional of finite type as a primitive notion; (3) archive material in which Gödel describes the Dialectica Interpretation as based on an intuitionistic insight obtained by an application of phenomenology; (4) archive material around the notion of reductive proof and its relation to phenomenology; and, in an appendix, (5) archive material according to which Gödel anticipated autonomous transfinite progressions when writing his incompleteness paper. A short companion paper describes archive material documenting the influence of Leibniz on the revision of the Dialectica paper (van Atten forthcoming).

M. van Atten (✉)
Sciences, Normes, Décision (CNRS/Paris 4), 28 rue Serpente, 75006 Paris, France
e-mail: vanattenmark@gmail.com

J. Dubucs, M. Bourdeau (eds.), *Constructivity and Computability in Historical and Philosophical Perspective*, Logic, Epistemology, and the Unity of Science 34,
DOI 10.1007/978-94-017-9217-2_7

7.1 Personal Contacts

7.1.1 Gödel and Brouwer

According to Wang (1987, p. 80), 'it appears certain that Gödel must have heard the two lectures' that Brouwer gave in Vienna in 1928; and in fact in a letter to Menger of April 20, 1972, Gödel says he thinks it was at a lecture by Brouwer that he saw Wittgenstein (Gödel 2003b, p. 133). But it is not likely that on that occasion Brouwer and Gödel had much, or indeed any, personal contact. In a letter of January 19, 1967, to George Corner of the American Philosophical Society, who had solicited a biographical piece on the then recently deceased Brouwer,[1] Gödel wrote that 'I have seen Brouwer only on one occasion, in 1953, when he came to Princeton for a brief visit'[2]; this is consistent with the above if Gödel meant that he had never actually talked to Brouwer before 1953.

In 1975 or 1976, Gödel stated that the first time he studied any of Brouwer's works was 1940.[3] A letter to his brother Rudolf in Vienna of September 21, 1941[4] documents an attempt to buy two of Brouwer's publications, his dissertation *Over de Grondslagen der Wiskunde* (*On the Foundations of Mathematics*) and the collection of three articles *Wiskunde, Waarheid, Werkelijkheid* (*Mathematics, Truth, Reality*).[5] The attempt is of some additional interest because of the situation in which it was written:

> Now I have a big favour to ask: could you order the following two books by L.E.J. Brouwer for me at Antiquarium K.F. Koehler (Leipzig, Täubchenweg 21)? 1. Over de Grondslagen der Wiskunde Katalog 115 No 487 2. Wiskunde, Waarheid, Werkelijkheid Groningen 1919. They are small books, which will cost only a few Marks. I am told that German bookstores ship books to foreign addresses without ado (probably at the risk of the recipient), if they

[1]December 2, 1966.

[2]Carbon copy in the Kurt Gödel Papers, Princeton, box 4c, folder 64, item 021257; more on that visit below. In the following, references of the form 'x/y, item n' are to item n in box x, folder y of the Kurt Gödel Papers. The documents in the microfilm edition of the archive do not always show an item number.

[3]See Gödel's draft replies to Grandjean's questionnaire, Gödel (2003a, pp. 447, 449); also Wang (1987, pp. 17, 19).

[4]Hence, after his Princeton lecture course on intuitionism of Spring 1941; according to Gödel's letter to Bernays of February 6, 1957, these were held at the Institute for Advanced Study (i.e., not in the Mathematics Department) (Gödel 2003a, p. 144). IAS Bulletin no. 9, of April 1940, gives as dates for the Spring Term of the academic year 1940–1941 February 1 to May 1. In the letter to Bernays, Gödel mentions that there exists no transcript of the course. However, his own lecture notes still exist, and are kept in the archive in 8c/121 (item 040407) and 8c/122 (item 040408). There are related notes in 8c/123, item 040409. Also the notes in 6a/54, item 030077, 'Beweis d[er] Gültigkeit d[er] int[uitionistischen] Ax[iomen]' belong with these.

[5]Brouwer (1907, 1919). The latter is a combined reprint of Brouwer (1908, 1909, 1912). The one place in Brouwer's papers between 1919 and 1941 where Brouwer (1907, 1919) are referred to together is footnote 1 of Brouwer (1922) (and its Dutch version Brouwer 1921).

> are ordered and paid for by a resident. On the other hand, from here nothing can be ordered from Germany through bookstores. Of course I make this order only in case the books are in stock. To have them searched for would come too expensive.[6]

The correspondence of the two brothers was then interrupted by the World War. Were the books sent? They are not in Gödel's personal library (although they might have disappeared from it). In the letters after the war, Gödel did not repeat the request.[7] Be that as it may, Gödel did at some point before 1952 make a detailed study of *Over de Grondslagen der Wiskunde*, as witnessed by the 13 pages of reading notes in his archive (10a/39, item 050135; on the envelope Gödel wrote '$< \overline{52}$').

When Brouwer visited Princeton in 1953, Gödel invited him twice: once for lunch and once for tea. From Gödel's remarks in a letter to his mother dated October 31, 1953 (Schimanovich-Galidescu 2002, p. 197), one gathers that Gödel did this because he felt obliged to. Indeed, Kreisel (1987b, p. 146) reports that 'Gödel was utterly bored by Brouwer', in spite of the latter's 'probably genuine exuberance'. Brouwer, in turn, in a letter to Morse of January 4, 1955 (van Dalen 2011, p. 455), sent his best wishes to several named people at the Institute, but did not include Gödel. A more positive, though less direct, connection between Brouwer and Gödel is that the author of the monumental handbook on English grammar Poutsma (1914–1929) that Gödel considered authoritative (Gödel 2003b, p. 303) was a maternal uncle of Brouwer's.

7.1.2 Gödel and Heyting

Gödel began to correspond with Brouwer's former student and then foremost follower, Arend Heyting, immediately after the Königsberg conference in 1930 that they had both attended. Like Brouwer, Heyting will not have been surprised by the incompleteness of formal systems for arithmetic, but Heyting acknowledged more explicitly the work behind it. Plans in the early 1930s for a joint book by Heyting and Gödel, which was to present an overview of contemporary research

---

[6]Wienbibliothek im Rathaus, Gödel Sammlung, item LQH0236598. Translation MvA. 'Jetzt habe ich noch eine grosse Bitte an Dich: Könntest Du die folgenden beiden Bücher von L.E.J. Brouwer beim Antiquarium K.F. Koehler (Leipzig, Täubchenweg 21) für mich bestellen? 1. Over de Grondslagen der Wiskunde Katalog 115 No 487 2. Wiskunde, Waarheid, Werkelijkheid Groningen 1919. Es sind kleine Bücher, die bloss ein paar Mark kosten werden. Man sagt mir dass Deutsche Buchhandlungen ohne weiteres Bücher an ausländische Adressen (wahrscheinlich auf Gefahr des Empfängers) versenden, wenn Sie von einem Inländer bestellt u. bezahlt werden. Andrerseits kann man von hier aus durch Buchhandlungen nichts aus Deutschland bestellen. Natürlich mache ich die Bestellung bloss für den Fall, dass die Bücher vorrätig sind. Sie suchen zu lassen käme zu teuer.'

[7]In a letter of August 3, 1947 (Wienbibliothek im Rathaus, Gödel Sammlung, item LQH0237199), he does ask Rudolf to find out in a bookstore whether anything had been published since 1941 by or about Leibniz.

in the foundations of mathematics, never quite materialized. Eventually, Heyting published his part separately (Heyting 1934), and Gödel never completed his.[8]

In December 1957, Gödel and Heyting met again. The occasion was a lecture tour that Heyting was making from the East to the West coast of the United States. In Princeton, Heyting gave two lectures at the Institute for Advanced Study, on Gödel's invitation,[9] and one at Princeton University. The titles and dates were 'Intuitionistic theory of measure and integration', IAS, December 9; 'The interpretation of intuitionistic logic', IAS, December 10; and 'On the fundamental ideas of intuitionism', Princeton University, December 11.[10]

William Howard was at the lecture on measure theory and integration, and recalls:

> I was working at Bell Labs at the time and it was only a short drive to Princeton. Nerode mentioned that Heyting was going to give a lecture, so we went. Gödel sat in the back. Then, during the question period at the end of the lecture, he got up with his little notebook in hand and started reading out a series of questions (criticisms). The only one I remember was as follows: Can a proposition which is neither true nor false today become true tomorrow (i.e., if someone proves it tomorrow)? Gödel really did not like the idea that truth could vary from one day to the next. He really went after Heyting, who seemed to be rather taken aback.[11]

It is somewhat curious that Gödel should have chosen the lecture on measure theory rather than the one on the interpretation of logic, which of course he knew would be held the next day, to voice this particular criticism; perhaps Gödel wanted to make his opening shot at the earliest possible occasion.

In that lecture on logic, Heyting restricted himself to discussing what he calls the 'originally intended interpretation', i.e., what has become known as the Proof Interpretation, but should really be called Proof Explanation[12]; in particular, he did not discuss Gödel's work from the 1930s.

Gödel, in his invitation letter, had expressed the hope that Heyting would 'be able to stay in Princeton for some days in addition to those when you will be giving the lectures so that we may discuss foundational questions with you'.[13] Because of

---

[8]See the Gödel-Heyting correspondence, and Charles Parsons' introduction to it, in Gödel (2003b). Draft notes by Gödel for this joint project are in 7a/10, item 040019.

[9]Gödel to Heyting, October 7, 1957, Heyting archive, item V57E-b-6.

[10]The texts of Heyting's lectures are held in the Heyting papers at the Rijksarchief Noord-Holland in Haarlem, items V57 (Dec 11), V57A (Dec 10), and V57B (Dec 9).

[11]Personal communication from William Howard, email to MvA, January 25, 2013.

[12]As Sundholm (1983, p. 159) points out, in logical-mathematical contexts, 'interpretation' has come to refer to the interpretation of one formal theory in another. In contrast, the so-called Proof Interpretation (also known as BHK-Interpretation) is not an interpretation in this mathematical sense, but a meaning explanation. Gödel's Dialectica Interpretation, on the other hand, indeed is one. Note that this immediately shows that the Proof Explanation and the Dialectica Interpretation differ in kind. Of course, a mathematical interpretation may devised because one has a particular meaning explanation in mind for the formulas it yields; this was Gödel's foundational aim with the Dialectica Interpretation.

[13]Gödel to Heyting, October 7, 1957, Heyting archive, item V57E-b-6.

Heyting's further commitments, he actually had to leave Princeton immediately. But according to a little diary that he kept of his American tour,[14] he did have an otherwise unspecified discussion with Gödel on December 6 or 7, lunch with Gödel on December 8, and a discussion with Gödel on impredicative definitions on December 9.

The latter discussion may well have touched on the clause for implication in Heyting's Proof Explanation, but no notes on its actual contents seem to exist.[15] It may also have included Gödel's distinction between 'predicative intuitionism' and 'impredicative intuitionism', which later led Kreisel, Myhill and others to develop the latter (Kreisel 1968, section 5; Myhill 1968, p. 175). Perhaps they talked about their shared conviction that constructive mathematics (understood as a foundational program) is not contained in classical mathematics, and is an altogether different subject (see page 187 below). Another possible topic in this conversation may have been Markov's Principle. Not long before Heyting's visit, Gödel had shown that if one can establish completeness of intuitionistic predicate logic relative to a so-called internal interpretation, this entails the validity of Markov's Principle.[16] Markov's Principle is rejected by most, though not all, intuitionists. Gödel's argument, which also goes through for the notions of validity defined by Beth and Kripke, therefore seems to show that the intuitionist cannot hope ever to establish completeness of intuitionistic predicate logic. This result was a motivation for Veldman and de Swart to develop alternative semantics, relative to which completeness does not entail Markov's Principle (de Swart 1976b, 1977; Veldman 1976). They treat negation not as absence of models but as arriving at a falsehood. In particular, de Swart (1977) presented a semantics that, within the limits of formalization, seems to mirror Brouwer's conception of mathematical activity quite faithfully.

The last contact between Heyting and Gödel seems to have been in 1969, when Heyting inquired if Gödel were interested, as was rumoured, in publishing his collected works. If true, Heyting continued, he would very much like to have them appear in the series *Studies in Logic* (North-Holland), of which he was one of the editors. But Gödel replied that he actually had no such interest, and that he considered such a project not very useful, as his important papers were all readily available (Gödel 2003b, pp. 74–75).

---

[14] Heyting archive, item V57E-r.

[15] William Howard (email to MvA, February 1, 2013) recalls:

> Re the issue of impredicativity in BHK: Gödel and I did not discuss this issue explicitly, but it was implicit in some of our discussions of my little theory of constructions (the formulae-as-types paper [Howard 1980], which then existed in the form of handwritten document, which I had sent to Gödel as part of my application for my sabbatical at the IAS, 1972–1973). Yes, he had obviously read the little paper.

[16] This result was published by Kreisel (1962, p. 142), who specifies that Gödel had obtained it in 1957. See for a discussion of the notion of internal validity Dummett (2000), section 5.6.

7.2 Philosophical Contacts

Gödel recognized the epistemological advantages of constructivism, and looked for interpretations of formal systems for intuitionistic logic and arithmetic. A characteristic feature of Gödel's technical results in this area is that none of them is concerned with the intuitionists' intended interpretations, except, perhaps, in the negative sense of avoiding them. To Sue Toledo, he said (at some point in the period 1972–1975) that 'intuitionism involves [an] extra-mathematical element. Namely, the mind of the mathematician + his ego', and he described intuitionism to her as 'essential a priori psychology' (van Atten and Kennedy 2009, p. 496). (I will come back to that characterization below.) This did not keep Gödel from studying specifically Brouwerian topics closely[17]; moreover, on various occasions Gödel has shown that he knew how to let (ideological) intuitionism inspire him in his own work. These will be commented on, or, in the last case, discussed at length, in the following sections:

- The Incompleteness Theorem (Sect. 7.2.1)
- Weak Counterexamples (Sect. 7.2.2)
- Intuitionistic Logic as a Modal Logic (Sect. 7.2.3)
- Continuity Arguments in Set Theory (Sect. 7.2.4)
- Around the Dialectica Interpretation (Sect. 7.2.5)

7.2.1 The Incompleteness Theorem

According to an entry in Carnap's diary for December 23, 1929,[18] Gödel talked to him that day

> about the inexhaustibility of mathematics (see separate sheet). He was stimulated to this idea by Brouwer's Vienna lecture. Mathematics is not completely formalizable. He appears to be right. (Wang 1987, p. 84)[19]

On the 'separate sheet', Carnap wrote down what Gödel had told him:

> We admit as legitimate mathematics certain reflections on the grammar of a language that concerns the empirical. If one seeks to formalize such a mathematics, then with each formalization there are problems, which one can understand and express in ordinary

---

[17] See the index to his Arbeitshefte (5c/12, item 030016) and the headings in the Arbeitshefte, both published in English in Dawson and Dawson (2005, pp. 156–168), as well as the remarks on Gödel and Brouwer's Bar Theorem further down in the present paper, and footnote 114.

[18] In line 14 on p. 498 of van Atten and Kennedy (2009), read '23' for '12'.

[19] 5 3/4–8 1/2 Uhr Gödel. Über Unerschöpflichkeit der Mathematik (siehe besonderes Blatt). Er ist durch Brouwers Wiener Vortrag zu diesen Gedanken angeregt worden. Die Mathematik ist nicht restlos formalisierbar. Er scheint recht zu haben. (Köhler 2002, p. 92)

> language, but cannot express in the given formalized language. It follows (Brouwer) that mathematics is inexhaustible: one must always again draw afresh from the 'fountain of intuition'. There is, therefore, no characteristica universalis for the *whole* mathematics, and no decision procedure for the whole mathematics. In each and every *closed language* there are only countably many expressions. The *continuum* appears only in 'the whole of mathematics' [...] If we have *only one language*, and can only make 'elucidations' about it, then these elucidations are inexhaustible, they always require some new intuition again. (Wang 1987, p. 50, trl. Wang?, original emphasis)[20]

Brouwer's argument in Vienna had been that no language with countably many expressions can exhaust the continuum, hence one always needs further appeals to intuition (Brouwer 1930, pp. 3, 6). Of course, the theorems that Gödel went on to demonstrate are of a different and much more specific nature.[21]

7.2.2 Weak Counterexamples

Also in Brouwer's Vienna lectures, Gödel will have noticed Brouwer's technique of the weak counterexamples Brouwer (1929). Gödel used this technique shortly after, and most effectively, when in 1930 he refuted Behmann's claim that classical existence proofs (not involving the uncountable infinite) can always be made constructive.[22]

7.2.3 Intuitionistic Logic as a Modal Logic

Another case is Gödel's translation of 1933 of intuitionistic propositional logic into the modal logic S4 (Gödel 1933b). Troelstra (Gödel 1986, p. 299) has pointed out that this translation was very likely inspired by Heyting's talk at

---

[20]'Wir lassen als legitime Mathematik gewisse Überlegungen über die Grammatik einer Sprache, die vom Empirischen spricht, zu. Wenn man eine solche Math. zu formulieren versucht, so gibt es bei jeder Formalisierung Probleme, die man einsichtig machen und in gewöhnlicher Wortsprache ausdrücken, aber nicht in der betroffenen formalisierten Sprache ausdrücken kann. Daraus folgt (Brouwer), dass die *Math. unerschöpflich* ist: man muss immer wieder von neuem aus dem "Born der Anschauung" schöpfen. Es gibt daher keine Characteristica universalis für die *gesamte* Math., und kein Entscheidungsverfahren für die gesamte Math. In irgend einer *abgeschlossenen Sprache* gibt es nur abzählbar viele Ausdrücke. Das *Kontinuum* tritt nur in der "gesamten Math." auf. [...] Wenn wir *nur eine Sprache* haben, und über sie nur "*Erläuterungen*" machen können, so sind diese Erläuterungen unausschöpfbar, sie bedürfen immer wieder neuer Anschauung.' (Köhler 2002, p. 110, original emphasis.) Note that Köhler, unlike Wang, does not explicitly identify this as the 'separate sheet' mentioned in the diary note; but both give the same date for it.

[21]For Brouwer's reaction to the incompleteness theorems, the reader is referred to section 3.5 of the on-line article van Atten (2012).

[22]See Gödel (2003a, p. 17, 2003b, pp. 565–567), and Mancosu (2002).

the Königsberg conference, which Gödel attended and of which he reviewed the published version (Gödel 1932; Heyting 1931). Heyting introduced a provability operator, but chose not to develop its logic. As he explained, on the intuitionistic understanding of mathematical truth, an explicit provability operator is redundant. Gödel's idea of truth, of course, was different.

7.2.4 Continuity Arguments in Set Theory

A use of intuitionistic ideas that goes beyond the heuristic is found in Gödel's work in set theory. In conversation with Hao Wang, Gödel claimed, 'In 1942 I already had the independence of the axiom of choice [in finite type theory]. Some passage in Brouwer's work, I don't remember which, was the initial stimulus' (Wang 1996, p. 86).[23] One can see what idea of Brouwer's Gödel was probably referring to by consulting Gödel's Arbeitsheft 14 which contains his notes on the proof dated 'ca. Ende März 1942'. There, Gödel uses Brouwer's continuity principle for choice sequences to define a notion of 'intuitionistic truth' for propositions about infinite sequences.[24] The principle states that if to every choice sequence a natural number is assigned, then for each sequence this number is already determined by an initial segment. By 1942, Gödel may have seen it in Brouwer's papers 1918, 1924a, 1924b, and 1927; in the Gödel archive, item 050066 in 9b/13 contains shorthand notes to the latter two. Unfortunately I have not been able to determine the date of these notes.[25] Gödel also described to Wang the method he had used as 'related' to Cohen's (Wang 1996, p. 251). In Cohen's forcing, too, truth values for propositions about certain infinite objects (generic sets) are always already determined by information about a finite part of such an object. This is of course not to suggest that Gödel invented forcing before Cohen: much more than the idea of finite approximations is needed to arrive at that.[26]

[23]Gödel did not publish this result; he states his reasons in a letter to Church of September 29, 1966 (Gödel 2003a, pp. 372–373) and in a letter to Rautenberg of June 30, 1967 (Gödel 2003b, pp. 182–183).

[24]5c/26, item 030032. See, e.g., pp. 14–16.

[25]On the otherwise empty back, Gödel wrote 'Brouwer bar theorem'; that English term was introduced only in Brouwer (1954). But it is not excluded that Gödel made these notes before or in 1942 and then added that jotting on the back later.

[26]For a detailed analysis of the analogy between forcing and intuitionistic logic, see Fitting (1969). In fact, Cohen's development of forcing after his initial discovery was influenced by this analogy, when Dana Scott pointed out to him how it could be used to simplify his treatment of negation; see Scott's foreword to Bell (1985). Scott there also mentions the anticipation of forcing in Kreisel (1961).

7.2.5 *Around the Dialectica Interpretation*

By far the closest rapprochement of Gödel to intuitionism, however, is seen in the change over the years in Gödel's conception of constructivity. It would probably be one-sided to consider this change part of intuitionism's legacy on Gödel, yet it is inextricably intertwined with his ponderings on the Proof Explanation from the early 1930s onward. Moreover this change was such that, as we will see, it actually brought Gödel closer to the Proof Explanation that he otherwise always criticized.

7.2.5.1 Early Qualms About the Proof Interpretation

Gödel's qualms with Heyting's Proof Explanation seem to have arisen as soon as it was devised. The problem, as Gödel voices it in his Cambridge lecture in 1933, is that the clause for negation (more generally, the clause for implication) involves the notion of arbitrary intuitionistic proof in an essential way, and that this notion is too indeterminate. It does not comply with a condition that Gödel at the time posed on constructivity:

> Heyting's axioms concerning absurdity and similar notions [...] violate the principle [...] that the word 'any' can be applied only to those totalities for which we have a finite procedure for generating all their elements [...] The totality of all possible proofs certainly does not possess this character, and nevertheless the word 'any' is applied to this totality in Heyting's axioms [...] Totalities whose elements cannot be generated by a well-defined procedure are in some sense vague and indefinite as to their borders. And this objection applies particularly to the totality of intuitionistic proof because of the vagueness of the notion of constructivity. (Gödel 1933c, p. 53)

Gödel says—at this point— that a general notion of intuitionistic proof would only be constructively acceptable if it forms a totality that can be generated from below. An intuitionist might reply that this is the wrong demand to make. What matters to the intuitionist is that 'we recognize a proof when we see one' (Kreisel). The clause for implication (and hence that for negation) is not to be understood as quantifying over a totality of intuitionistic proofs—something that for a principled intuitionist like Brouwer or Heyting does not exist. Rather, the clause should be understood as expressing that one has a construction that, whenever a proof is produced that one recognizes as a proof of the antecedent, can be used to transform that proof into a proof of the consequent. Although an intuitionist believes the notion of proof to be open-ended, this understanding of implication can be expected to work because in proofs of implications usually nothing more is assumed about a proof of the antecedent than that it indeed is one.

The prime example of an intuitionistic theorem that goes beyond that assumption is Brouwer's proof of the Bar Theorem (Brouwer 1924a,b, 1927, 1954). This (classically trivial, but constructively remarkable) theorem basically says that, if a tree contains a subset of nodes such that every path through the tree meets it (a 'bar'), then there is a well-ordered subtree that contains a bar for the whole tree. Brouwer's extraction of additional information from the hypothesis that we have obtained a proof of the antecedent (i.e., that we have obtained a proof that the tree contains a bar) is based on his analysis of proofs as mental structures, and of mathematical objects as mentally constructed objects (a view wholly opposed to Gödel's). These analyses enable Brouwer to formulate a necessary condition for having a proof of the antecedent, namely, that it admit of being put into a certain canonical form (Brouwer 1927, p. 64); on the basis of that canonical form, a proof of the consequent is obtained. So what enables Brouwer to say something about all proofs of the antecedent is not the availability of a method that generates exactly these, but an insight into their mental construction that yields a necessary condition on them. In effect, Brouwer deals with 'such vast generalities as "any proof" ' by presenting a transcendental argument; for more on this, see Sundholm and van Atten (2008).[27]

In notes of 1938 for his 'Lecture at Zilsel's', Gödel states that 'Heyting's system [for intuitionistic arithmetic] violates all essential requirements on constructivity' (Gödel 1938b, p. 99). In his Yale lecture of April 15, 1941, 'In what sense is intuitionistic logic constructive?', he phrases his objection to Heyting's Proof Explanation by saying that the clause for implication requires that

> the notion of derivation or of proof must be taken in its intuitive meaning as something directly given by intuition, without any further explanation being necessary. This notion of an intuitionistically correct proof or constructive proof lacks the desirable precision. (Gödel 1941, p. 190)

Then Gödel goes on to present, in an informal manner, an interpretation of HA in a system of higher types Σ, and explains the motivations behind it. This use of functionals for a consistency proof of arithmetic is the main step forward compared to the discussion of functionals in the Lecture at Zilsel's. It seems Gödel had found the heuristic how to do this on January 1, 1941.[28] In the lecture, Gödel

---

[27]In a draft note for the revision of the Dialectica paper (9b/148.5, item 040498.59), Gödel wrote: 'Finally I wish to note that the definition of a proof as an unbroken chain of immediate evidences should be useful also for Heyting's interpretation of logic. In particular $A \supset B$ can then be defined simpler, namely by requiring that a proof of $A \supset B$ is a finite sequence P_i of propositions ending with B and such that each $P_i \neq A$ is immediately evident, either by itself, or on the basis of some of the preceding propositions.' A proof of the Bar Theorem based on that explanation of $\supset$ has not yet been found; compare Gödel's footnote d (Gödel 1990, p. 272).

[28]5c/19, Arbeitsheft 7 (item 030025), pp. 12–15 in the backward direction. This note is labeled 'Gentzen'. In the index to the Arbeitshefte (5c/12, item 030016), the reference to this note is the first entry under the heading 'Interpr.[ation] d.[er] int.[uitionistischen] Logik', and has '(heur.[istisch])' written after it. It contains, for example, a version of the proof for the validity of modus ponens interpreted by functionals. This must be the note that Wang describes, without a specific reference, in note 13 on p. 47 of Wang (1987). The date '1./I.1941' is on top of p. 12 (backward direction) of Arbeitsheft 7. On the same page, before the note 'Gentzen', there is one on 'Rosser Wid[erspruchs]fr [eiheits] Bew[eis]', with a horizontal line in between; that the date also holds for the second item on the page is very likely because it is also the date of another note, headed 'Jede F[u]nct[ion] d[es] eigentl[ich] intuit[ionistischen] Systems ist berechenbar'. That note is in the notebook *Resultate Grundlagen III* (6c/85, item 030118, pp. 188–191) and states the date '1./I.1941'. It begins with a reference to p. 34 (backward direction) of Arbeitsheft 7, which is where the formal system Σ is defined.

proposes schemata for defining the constructive operations used in the definition of computable function. He admits, however, that 'a closer examination of the question in which manner the functions obtained by these two schemes are really calculable is pretty complicated' (Gödel 1941, p. 195); the formal proof he had depends on Heyting arithmetic and hence is, foundationally speaking, no progress.

In the Dialectica paper of 1958, Gödel decides to take the notion of computable functional as primitive. Philosophically, this marks a sea change, which is discussed in the next subsection.

7.2.5.2 The Shift to the Intensional

Kreisel writes that 'In the first 20 min of our first meeting, in October 1955, [Gödel] sketched some formal work he had done in the forties, and later incorporated in the socalled Dialectica interpretation (with a total shift of emphasis)' (Kreisel 1987b, p. 104). I take the 'shift of emphasis' to be what I here call the sea change. (Note also that on p. 110, Kreisel says that, when he saw the Dialectica paper in 1958, 'the principal novelty—both absolutely speaking, and to me personally' was 'the primitive notion of effective rule [. . .] Gödel had never breathed a word to me about his project of exploiting such a notion'.) I therefore do not agree with Feferman who comments on Kreisel's report that 'Evidently Gödel misremembered: there is really no significant difference in emphasis, though the 1941 lecture mentions a few applications that are not contained in the 1958 Dialectica article' (Feferman 1993, p. 220).

In the Dialectica paper, Gödel defends his new approach by pointing to a similar case:

> As is well known, A.M. Turing, using the notion of a computing machine, gave a definition of the notion of computable function of the first order. But, had this notion not already been intelligible, the question whether Turing's definition is adequate would be meaningless. (Gödel 1990, p. 245n. 6)[29]

Mutatis mutandis, Gödel could have written this about his former self. He could have written: 'In 1941 I tried to give a definition of the constructive operations used in the definition of computable functional of finite type. But, had this notion not already been intelligible, the question whether my definition of 1941 is adequate would be meaningless.' In Gödel's view, what Turing had done was to define (and in that sense see sharper) an objective concept that we had been perceiving all along, albeit less sharply (Wang 1974, pp. 84–85). Similarly, Gödel holds in 1958, there is an objective concept of computable functional of finite type, which we may not

[29] A.M. Turing hat bekanntlich mit Hilfe des Begriffs einer Rechenmaschine eine Definition des Begriffs einer berechenbaren Funktion erster Stufe gegeben. Aber wenn dieser Begriff nicht schon vorher verständlich gewesen wäre, hätte die Frage, ob die Turingsche Definition adäquat ist, keinen Sinn. (Gödel 1958, p. 283n. 2).

yet (and possibly never) be able to make completely explicit, but which at the same time we see enough of to determine some of its properties:

> One may doubt whether we have a sufficiently clear idea of the content of this notion [of computable functional of finite type], but not that the axioms given [in this paper] hold for it. The same apparently paradoxical situation also obtains for the notion, basic to intuitionistic logic, of a proof that is informally understood to be correct. (Gödel 1990, p. 245n. 5)[30,31]

The point Gödel makes in this footnote is reminiscent of paragraph XXIV in Leibniz' *Discours de Métaphysique* (Leibniz 1875–1890, vol. 4, p. 449), where attention is drawn to the fact that there are situations in which we are able to classify certain things correctly and perhaps moreover explain the grounds on which we do this, yet without having at our disposal a complete analysis of the notion of those things into primitive terms. Whether Gödel, who surely knew that passage at the time,[32] also had it in mind when writing his footnote, remains an open question.[33]

Gödel's point here goes well with the disappearance in 1958 of his earlier denial that the Proof Explanation is genuinely constructive. Indeed, Gödel's own earlier objections to the intuitionistic notion of proof would equally apply to the primitive notion he substitutes for it in 1958. Since he considers the latter to be constructive (given his by then widened notion of constructivity), these earlier objections could no longer be used to support a claim that the Proof Explanation is not at all constructive; and the 1958 paper proposes no replacements for them. The difference has become one of degree, not of kind.[34]

---

[30] Man kann darüber im Zweifel sein, ob wir eine genügend deutliche Vorstellung vom Inhalt dieses Begriffs haben, aber nicht darüber, ob die weiter unten angegebenen Axiome für ihn gelten. Derselbe scheinbar paradoxe Sachverhalt besteht auch für den der intuitionistischen Logik zugrunde liegenden Begriff des inhaltlich richtigen Beweises. (Gödel 1958, p. 283n. 1).

[31] Compare Gödel's claim, in the Lecture at Zilsel's from 1938, that the axioms of the subsystem of Heyting's logic presented there are, when interpreted intuitionistically, 'actually plausible' ('tatsächlich plausibel'; Gödel 1938b, p. 100/101).

[32] Notes he wrote to it when reading it in Gerhardt's edition can be found in folder 10a/35. As Sundholm reminded me, the same distinctions are explained by Leibniz also in an earlier text of the same period (published in the same volume of Gerhardt's edition), the 'Meditationes de cognitione, veritate, et ideis' of 1684 (Leibniz 1875–1890, vol. 4, pp. 422–426). That text actually served as a basis for the section in the *Discours*. Leibniz takes up the theme again in the *Nouveaux Essais* of 1704 (but published posthumously, in 1765), Book II, chapter XXXI, § 1 (Leibniz 1875–1890, vol. 5, pp. 247–248).

[33] There is a direct and documented relation between ideas of Leibniz and the revisions of the Dialectica Interpretation; see van Atten (forthcoming).

[34] This is most explicit in the 1972 version, for the more specific interpretation in terms of reductive proof, which Gödel says is 'constructive and evident in a higher degree than Heyting's' (Gödel 1972a, p. 276n.h).

What might have been Gödel's reason for changing his tune on the conditions of constructivity? And when did he do this? Two as yet unpublished sources that are relevant here are Gödel's notes for the Princeton lecture course on intuitionism of Spring 1941 and his philosophical notebook Max IV (May 1941 to April 1942).[35] From a letter to his brother Rudolf of March 16, 1941, we know that Gödel worked on the Princeton lecture course and the Yale lecture at the same time:

> Now I have again many things to do, because I give a lecture course and in addition have again[36] been invited to give a lecture, where in both cases the topic is my most recent work, which I haven't put to paper in exact form yet even for myself.[37]

(By the end of the lecture course, there were only three students left[38]; it would be interesting to know who they were.) In his notes for the lecture course, Gödel writes, when he arrives at the concrete question of the computability of the functionals of his system Σ:

> I don't want to give this proof in more detail because it is of no great value for our purpose for the following reason. If you analyse this proof it turns out that it makes use of logical axioms, also for expressions containing quantif[iers] and since [sic] it is exactly these ax[ioms] that we want to deduce from the system Σ. [8c/122, item 040408, p. 61; also Gödel (1995, p. 188)]

Then follow two alternative continuations for this passage; here labelled (A) and (B). (B) is also quoted in volume III of the *Collected Works*, but (A) is not. (A) is written immediately below the previous quotation and reads:

> (A) So our attitude must be this that the ax[ioms] of Σ (in part[icular] the schemes of def[inition]) must be admitted as constructive without proof and it is shown that the ax[ioms] of int[uitionistic] logics[39] can be deduced from them with suitable def[initions]. This so it seems to meis a progress [8c/122, item 040408, pp. 61–62]

Gödel crossed out (A). (B) follows immediately after it, and is not crossed out:

> (B) There exists however another proof. Namely it is possible instead of making use of the log[ical] operators applied to quantified expression[s] to use the calculus of the ordinal nu[mbers] (to be more exact of the ord[inal] nu[mbers] $< \varepsilon_0$). I shall speak about this proof later on. [8c/122, item 040408, p. 62; Gödel (1995, p. 189)]

---

[35]For the archive numbers of the lecture course, see footnote 4 above. The notebook is in 6b/67, item 030090.

[36][On November 15, 1940, Gödel had lectured at Brown University on the consistency of the Continuum Hypothesis.]

[37] Wienbibliothek im Rathaus, Gödel Sammlung, item LQH0236556. 'Ich habe jetzt wi[eder] eine Menge zu tun, da ich eine Vorlesung halte u[nd] ausserdem wieder zu einem Vortrag eingeladen bin wobei in beiden Fällen das Thema meine allerletzte Arbeiten sind, die ich noch nicht einmal für mich selbst genau zu Papier gebracht habe.'

[38]Letter to Rudolf Gödel, May 4, 1941 (Wienbibliothek im Rathaus, Gödel Sammlung, item LQH0236557): 'Hier ist jetzt das Semester zu Ende u[nd] ich bin froh dass mit meiner Vorlesung Schluss ist, ich hatte zum Schluss nur mehr 3 Hörer übrig.' As mentioned in footnote 4 above, the Spring Term had ended on May 1.

[39][Perhaps Gödel uses the plural here because he is thinking of intuitionistic logic as it figures in different theories.]

Gödel then introduces (pp. 62 and 63) the idea of an ordinal assignment to terms such that with a reduction of a term comes a decrease of the ordinal. In an alternative version of (B), on p. 63^{iii}, here labelled (C), Gödel writes:

> (C) However it seems to be possible to give another proof which makes use of transfinite induct[ion] up to certain ord[inals] (probably up to [the] first ε-nu[mber] would be sufficient). [8c/122, item 040408, p. 63^{iii}; Gödel (1995, p. 189)]

It seems that Gödel wrote (A) before (B) and (C) and that he preferred the possible solution described in the latter two. This preference for (B) and (C) however does not seem to indicate a categorical rejection of (A), for on p. 63^{iv}, which follows right after (C), Gödel goes on to comment:

> (D) Of course if you choose this course then the question arises in which manner to justify the inductive inference up to the certain ordinal nu[mber] and one may perhaps be of the opinion that the ax[ioms] of Σ are simpler as a basis than this transfinite m[ethod], by [...][40] to justify them. But whatever the opinion to this question may be in any case it can be shown that int[uitionistic] logic if applied to nu[mber] theory (and also if applied in this whole system Σ) can be reduced to this system Σ. [8c/122, item 040408, p. 63^{iv}]

(Like (A), this passage is not quoted in the *Collected Works*.)

On the one hand, as Troelstra remarks on (B) and (C), 'Since the notes do not contain any further particulars, it is not likely that Gödel had actually carried out such a proof in detail' (Gödel 1995, p. 189).[41] On the other hand, (A), even crossed out, and the somewhat less emphatic (D), show that Gödel already around the time of the Yale lecture, in which there is no mention of the possibility of accepting the notion of computable functional as primitive, had considered doing just that.

Gödel's philosophical notebook Max IV, which covers the period from May 1941 to April 1942, that is, the period immediately after the Princeton course and the Yale lecture, contains the following remark:

> Perhaps the reason why no progress is made in mathematics (and there are so many unsolved problems), is that one confines oneself to ext[ensions]—thence also the feeling of disappointment in the case of many theories, e.g., propositional logic and formalisation altogether.[42]

[40]Two or three words that are difficult to read; perhaps 'which we try'?

[41]I have not attempted to reconstruct, from the Arbeitshefte, how far Gödel got. But he evidently did not succeed: in conversation with Kreisel in 1955, he mentioned the assignment of ordinals as an open problem (Kreisel 1987b, p. 106), and, although it was solved for a special (but in a sense sufficient) case in Howard (1970), he did so again in a telephone conversation with Tait in 1974. But, as Tait remarks, to exploit such an assignment in a proof of normalization, PRA together with induction up to ε_0 are required, so it could not serve Gödel's foundational aim (Tait 2001, p. 116 and its n. 39). (See Kanckos (2010) for a version of Howard's proof in the setting of Natural Deduction.)

[42]'Vielleicht kommt man in der Math[ematik] deswegen nicht weiter (und gibt es so viele ungelöste Probl[eme]), weil man sich auf Ext[ensionen] beschränkt—daher auch das Gefühl der Enttäuschung bei manchen Theorien, z.B. dem Aussagenkalkül und der Formalisierung überhaupt.' 6b/67, item 030090 (Notebook Max IV), p. 198. Transcription Cheryl Dawson and Robin Rollinger.

The disappointment in the case of propositional logic that Gödel speaks of here may well be a reference to the fact that the difficulties he ran into when attempting a (foundationally satisfying) formal reconstruction of intuitionistic logic within number theory in the system Σ of the Yale lecture appeared already with the propositional connectives; his disappointment with formalization (an act that pushes one to the extensional view) in general may have found additional motivation in his incompleteness theorems.

This remark from 1941–1942, with its implicit recommendation to shift emphasis to the intensional, strongly suggests that by then Gödel had indeed come to accept the solution proposed in passage (A) of the Princeton lecture course. In print, this would become clear of course only in 1958. Perhaps this view on the development of the Dialectica Interpretation would need some refinement in light of Kreisel's report that

> Gödel made a point of warning me [in 1955] that he had not given any thought to the objects meant by (his) terms of finite type. The only interpretation he had in mind was formal, as computation rules obtained when the equations are read from left to right. (Kreisel 1987a, p. 106).

But in the light of (A) and (D), it seems to me that the claim ascribed to Gödel in the first sentence here cannot be quite correct; unfortunately, we do not have Gödel's own words. (In the notes for the Princeton lectures, Gödel also defined and used a model in terms of what became known as the hereditarily effective operations (HEO)[43]; but unlike the primitive notion of computable functionals and the method of assigining ordinals, HEO has, because of the logic in its definition, no significance for the foundational aim that Gödel hoped to achieve.)

Before continuing the discussion of the shift to the intensional, it is worth noting that, as for the purely proof-theoretical applications of the interpretation described in the Yale lecture, according to Kreisel Gödel 'dropped the project after he learnt of recursive realizability that Kleene found soon afterwards' (Kreisel 1987a, p. 104). (Kleene told Gödel about realizability in the summer of 1941 (Kleene 1987, pp. 57–58).) In contrast to realizability, the functional interpretation lends itself to an attempt to make the constructivity of intuitionistic logic (within arithmetic) more evident, and, as I have tried to show, that was Gödel's purpose for it already by 1941 or 1942. This raises the question why Gödel waited until 1958 to publish these ideas. In an undated draft letter to Frederick W. Sawyer, III (written after February 1, 1974, the date of Sawyer's letter to which it is a reply), Gödel says:

> It is true that I first presented the content of my Dialectica paper in a course of lectures at the Institute in Spring   and in a talk at Yale in   . There were several reasons why I did not publish it then. One was that my interest shifted to other problems, another that there was not too much interest in Hilbert's Program at that time. (Gödel 2003b, p. 211; spaces left open by Gödel)[44]

[43] 8c/123, item 040409, p. 109ff; see also Gödel (1995, pp. 187–188).

[44] At the beginning of the Yale lecture, Gödel said that 'the subject I have chosen is perhaps a little out of fashion now' (Gödel 1995, p. 189); and he told Wang in April 1977 that at the Yale lecture, 'nobody was interested' (Wang 1996, p. 86).

The shift to the intensional had its first effect in print soon, in Gödel's remarks on analyticity in the Russell paper of 1944 (Gödel 1944, pp. 150–153), in particular in the last sentence of its footnote 47:

> It is to be noted that this view about analyticity [i.e., truth owing to the meaning of the concepts] makes it again possible that every mathematical proposition could perhaps be reduced to a special case of $a = a$, namely if the reduction is effected not in virtue of the definitions of the terms occurring, but in virtue of their meaning, which can never be completely expressed in a set of formal rules.

There are a number of later echoes of the remark of 1941 or 1942 in Gödel's writings, published and unpublished. I mention three that are directly related to Gödel's development of the Dialectica intepretation. The first is Kreisel's report that, when in October 1955 Gödel explained the formal part of the functional interpretation to him, Gödel added a warning about the 'Aussichtlosigkeit, that is, hopelessness of doing anything decisive in foundations by means of mathematical logic' (Kreisel 1987a, p. 107; p. 104 for the date). The relation between (existing) mathematical logic and extensionality that one must see in order to connect the remark of 1941 and this warning is made explicit in the second echo. It occurs in Gödel's letter to Bernays of July 14, 1970, concerning the revision of the Dialectica paper: 'The mathematicians will probably raise objections against that [i.e., the decidability of intensional equations between functions], because contemporary mathematics is thoroughly extensional and hence no clear notions of intensions have been developed.' (Gödel 2003a, p. 283). The third echo also has to do with that revision, and is a draft for part of note k (the later note h in the *Collected Works* (Gödel 1990, pp. 275–276), but in the version published there Gödel had decided not to include this):

> This note (and also some other parts of this paper) constitutes a piece of 'meaning analysis', a branch of math[ematical] logic which, although it was its very starting point, today is badly neglected in comparison to the purely math[ematical] branch which has developed amazingly in the past few decades. (The reason of this phenomenon doubtless is the antiphil[osophical] attitude of today's science.) [9b/145][45]

Indeed, Kreisel has observed that, compared to the 1930s, 'later Gödel became supersensitive about differences in meaning'. He illustrated this, appropriately, by the contrast in attitude between Gödel's remark in 1933 that intuitionistic arithmetic involves only 'a somewhat deviant interpretation' from its classical counterpart and the caveat in 1958 that 'further investigation is needed' to determine to what extent the Dialectica Interpretation can replace the intuitionistic meanings (see also footnote 27 above).[46]

---

[45]Compare also Gödel to Bernays, September 30, 1958: 'Kreisel told me that in your lectures in England you discussed the combinatorial concept of set in detail. I very much regret that nothing about that has appeared in print. Conceptual investigations of that sort are extremely rare today.' (Gödel 2003a, p. 157)

[46]Gödel (1933a), p. 37/ 1986, p. 295; Gödel (1958), p. 286/ 1990, p. 251; Kreisel (1987b), pp. 82, 104–105, 159.

The shift in Gödel's view described here constitutes a remarkable rapprochement with intuitionism, which by its very nature takes intensional aspects to be the fundamental ones in mathematics. On both Gödel's new view and the intuitionistic one, foundational progress will therefore have to come mainly from informal analysis of intuitive concepts. To that end, Gödel around 1959 made an explicit turn to phenomenology as a method (van Atten and Kennedy 2003; Gödel 1961/?),[47] and encouraged Kreisel's developing and advocating the notion of informal rigour (Kreisel 1967) (at the same time warning him that mathematicians would not be enamoured of the idea).[48] Brouwer did not make an explicit turn to phenomenology, but his work lends itself to phenomenological reconstruction (van Atten 2004, 2006, 2010). They of course differed (in effect) on *which* phenomenological aspects are relevant to pure mathematics, because they had different conceptions of what pure mathematics, as a theory, consists in. Both took it to give an ontological description, but where for Gödel the domain described is a platonist realm, for Brouwer it is that of our mental constructions.[49] A telling anecdote related to this was related to me by William Howard:

> I don't remember the context, but I started to talk about 'Brouwer's bar theorem', for which Brouwer gave a sort of justification but certainly not a proof. As soon as I got the words 'Brouwer's bar theorem' out of my mouth, Gödel interrupted me, saying,
> 'But he did not provide a PROOF!'.[50]
> This was delivered with strong emotion and quite aggressively. I sat there thinking: well, it's not *my* fault. But I replied,
> 'Well, yes, I agree,'
> and then went on with whatever I had been saying.
> [original emphasis][51]

[47]In a letter to Gödel of June 17, 1960, written after a visit to him, Sigekatu Kuroda wrote: 'It was my great pleasure also that I heard from you that you are studying Husserl and you admired his philosophy, which was the unique philosophy that I devoted rather long period and effort in my youth. I hope I have a chance some day to speak with you about Husserl. As you are doing now, I would like to recollect Husserl's philosophy after returning to my country.' (01/99, item 011378) Note that by that time Kuroda had published philosophical and technical work on intuitionistic logic, notably Kuroda (1951), in which he moreover says (p. 36) that he shares Brouwer's view that mathematics is an activity of thought that is independent of logic and based on immediate evidence that is intuitively clear. Without further sources it is of course impossible to tell whether Gödel and Kuroda discussed phenomenology and intuitionism in relation to one another, but Kuroda's letter gives the impression that they had not.

[48]Personal communication from Georg Kreisel, letter to MvA, January 10, 2005.

[49]For more on ontological descriptivism, Brouwer's exploitation of it, and its contrast to meaning-theoretical approaches to mathematics such as Dummett's or Martin-Löf's, see section 5 of Sundholm and van Atten (2008).

[50][Note MvA: See also Gödel (1972a), p. 272, note d: 'Unfortunately, however, no satisfactory constructivistic proof is known for either one of the two principles [i.e., Brouwer's bar induction and Spector's generalization to finite types]'.]

[51]Story 20, p. 177 of Howard, unpublished. In an email to me of January 25, 2012, Prof. Howard adds that this was the only occasion during his conversations with Gödel (which took place during Howard's year at the IAS, 1972–1973) that the topic of the bar theorem and of bar induction came up.

The drafts for the revised version of the Dialectica paper (those that so far have remained unpublished) shed further light on Gödel's views on intuitionism and relate them to his turn to phenomenology. There are four versions to consider (together with the notes Gödel wrote when working on these); see Table 7.1. As is to be expected from what for Gödel was work in progress, the drafts and notes D68–D72 are homogeneous in neither form nor content. Although written with an eye on publication, they can of course not be granted the same status as Gödel's published work. But various passages in them are coherently related to each other and to remarks Gödel has made elsewhere, and in any case allow one to document his thinking on these matters. For that reason, I should like to discuss a (necessarily, limited) selection of these materials, in particular from the new introduction in D68.[52] That new introduction exists in the archive as a set of pages in longhand numbered 1–26 and 1F–12F (for footnotes).[53] As Gödel wrote to Bernays on May 16, 1968 that it is 'essentially finished', and on December 17, 1968 that 'in the end I liked the new one as little as the old' (Gödel 2003a, pp. 261, 265), I will give 1968 as the date of these drafts.

Table 7.1 Revised versions of Gödel (1958)

D67	—	The translation of Gödel (1958) by Leo Boron (in collaboration with William Howard), revised by Gödel but without substantial additions[54]
D68	—	A version that is essentially D67 with a rewritten, longer philosophical introduction[55]
D70	—	A version that is essentially D67 with an additional series of notes a–m. Circulated on Gödel's request by Dana Scott. Galley proofs exist[56]
D72	—	A revised version of D70. Last version available, published in *Collected Works* as Gödel (1972a)[57]

[52]The material is rich, and should also be studied with other questions in mind, and from other perspectives. To my mind, in particular D68 would have deserved to be included in the *Collected Works* as well.

[53]539b/141, item 040450.

[54]9b/141, item 040449 and 9b/142, item 040451

[55]The material for the new introduction is 9b/141, item 040450.

[56]9b/144, item 040454 (I thank Dirk van Dalen for letting me photocopy the purple-ink duplicate he received from Scott in Oxford); 9b/149 for the galley proofs, items 040456 and 040459.

[57]See the material for D70 (and the revisions on it), and, for the revisions of notes c and k, 9b/145 items 040560 and 040457, respectively.

7.2.5.3 The Relation of Constructive Mathematics to Classical Mathematics

If constructive mathematics is conceived in such a way as to involve reference to properties of the mathematician's mental acts, then this explains Gödel's view, reported by Shen Yuting in a letter to Hao Wang of April 3, 1974, that 'classical mathematics does not "include" constructive mathematics' (3c/205, item 013133).[58] For a classical mathematician holding this view, considerations about, and results from, constructive mathematics (in the sense described) have mathematical significance only when they can be 'projected into the mathematically objective realm'. The choice of words here is Bernays', who uses it to define the distinction between a 'reserved' and a 'far-reaching' intuitionism in a letter of March 16, 1972 Gödel (2003a, pp. 295). Perhaps Gödel had a similar distinction in mind when, in a draft note for the revision of the Dialectica paper, he sees a need to give a characterization of finitism the form of a 'translation' of its traditional conception, as follows:

> Translating the definition of finitism given above into the language of modern mathematics (which does not consider spacetime intuition to belong to its field) one may say equivalently: The objects of finitary mathematics are hereditarily finite sets (i.e., sets obtained by iterated formation of finite sets beginning with a finite number of individuals or the 0-set); and finitary mathematics is what can be made evident about these sets and their properties, relations, and functions (definable in terms of the ∈-relation) without stepping outside this field of objects, and using from logic only propositional connectives, identity, and free variables for hereditarily finite sets. Clearly on this basis recursive definitions (proceeding by the "rank" of the sets) are admissibile as evidently defining well-determined functions without the use of bound variables. [...] [9b/141, item 040450, p. 2F; 1968]

It is, a fortiori, not surprising then that Gödel never came to accept in his own work on constructive mathematics the objects and techniques that are typical for Brouwer's 'far-reaching' intuitionism, such as choice sequences, Brouwer's proof of the Bar Theorem, and creating subject arguments.

7.2.5.4 Effective But Nonrecursive Functions

This distinction between a reserved and a far-reaching intuitionism is also important for the question whether there exist effective but nonrecursive functions.

This particular question had gained importance for Gödel by the time he came to revise the 1958 paper, as is clear from a comparison of the two versions of his footnote on Turing. While in 1958 he had written,

---

[58] This was also Heyting's view: 'I must protest against the assertion that intuitionism starts from definite, more or less arbitrary assumptions. Its subject, constructive mathematical thought, determines uniquely its premises and places it beside, not interior to classical mathematics, which studies another subject, whatever subject that may be' Heyting (1956, p. 4).

> As is well-known, A.M. Turing, using the notion of a computing machine, gave a definition of the notion of computable function of the first order. But, had this notion not already been intelligible, the question whether Turing's definition is adequate would be meaningless.[59] Gödel (1990, p. 245n. 6)[60]

in the 1972 version this became

> It is well known that A.M. Turing has given an elaborate definition of the concept of a *mechanically* computable function of natural numbers. This definition most certainly was not superfluous. However, if the term 'mechanically computable' had not had a clear, although unanalyzed, meaning before, the question as to whether Turing's definition is adequate would be meaningless, while it undoubtedly has an affirmative answer. (Gödel 1972a, p. 275n. 5, original emphasis)

In the latter version, the mechanical character of Turing's notion is made explicit and is emphasized. It might have been natural then also to ask about constructively evident but non-mechanical computability. Gödel chose not to do so in D72. He had in D68:

> In my opinion there are no sufficient reasons for expecting computability by thought procedures to have the same extension [as mechanical computability], in spite of what Turing says in Proc. Lond. Math. Soc. 42 (1936), p. 250. However, it must be admitted that, even in classical mathematics, the construction of a welldefined thought procedure which could actually be caried out and would yield a numbertheoretic function which is not mechanically computable would require a substantial advance in our understanding of the basic concepts of logic and mathematics and of our manner of conceiving them. [9b/141, item 040450, pp. 20–21; 1968]

In 1972, Gödel prepared a slightly different version of this remark for publication outside the Dialectica paper, Gödel (1972b), p. 306 (remark 3).[61]

However, in 'far-reaching' intuitionistic mathematics, Kripke has devised (but not published) an example of just such a function. The presentation I will follow

---

[59]Moreover, Gödel will have known the observation by Skolem, Heyting, and Péter that in constructivism, 'computable function' cannot be taken to *mean* 'recursive function'. See Skolem (1955, p. 584), a paper to which my attention was drawn by Thierry Coquand's contribution to the present volume; Heyting (1958, pp. 340–341), which appeared in the same special issue of *Dialectica* as Gödel's paper; Péter (1959). Heyting is the one who emphasizes the alternative of taking that notion as primitive. Tait (2006, pp. 212–213) holds that the fact that a definition would be circular shows that there is a problem with the idea of constructive evidence for the computability of a function. To my mind, that is not correct, but I will not develop this point here. See also Kreisel's review (1969) of Tait (1967).

[60] A.M. Turing hat bekanntlich mit Hilfe des Begriffs einer Rechenmaschine eine Definition des Begriffs einer berechenbaren Funktion erster Stufe gegeben. Aber wenn dieser Begriff nicht schon vorher verständlich gewesen wäre, hätte die Frage, ob die Turingsche Definition adäquat ist, keinen Sinn. (Gödel 1958, p. 283n. 2).

[61]Yet another version was published, with Gödel's approval, in Wang (1974), pp. 325–326 (reprinted in Gödel (2003b), p. 576).

here is that of van Dalen (1978), p. 40n. 3, which owes its elegance to its explicit use of the so-called Theory of the Creating Subject, CS (Kreisel 1967, pp. 159–160).[62] Write $\Box_n A$ for 'The creating subject has at time n a proof of proposition A'. Let K be a set that is r.e., but not recursive. Define

$$f(n,m) = \begin{cases} 0 \text{ if } & \Box_m n \notin K \\ 1 \text{ if not } & \Box_m n \notin K \end{cases}$$

For the creating subject, f is effectively computable, as at any given moment m, it is able to determine whether $\Box_m n \notin K$. By the standard principles governing the creating subject,[63] we have $n \notin K \leftrightarrow \exists m f(n,m) = 0$; this means that, if f were moreover recursive,[64] then the complement of K would be r.e., which contradicts the assumption.

According to van Dalen (in conversation), at the Summer Conference on Intuitionism and Proof Theory, SUNY at Buffalo, 1968, the example was considered common knowledge. Kreisel learned the example before that conference, perhaps from Kripke himself when the latter visited Stanford somewhere between 1963 and 1965.[65] Kreisel presented the theory CS to Gödel in a letter of July 6, 1965 (01/87, item 011182). I have not yet been able to determine yet whether Gödel came to know Kripke's function as well. But it seems likely that he did, given his close contact with Kreisel at the time.

[62]It is also possible to avoid CS, formally, by using the Brouwer-Kripke Schema BKS instead, usually formulated as $\exists \alpha (\exists n \alpha(n) = 1 \leftrightarrow A)$ (but the parenthetical qualification in the following footnote also holds here: BKS should really be formulated as two rules with parameters P and $\alpha = \alpha_P$). However, from the intuitionistic point of view, the known justification of BKS also justifies CS. Versions using BKS were given by Gielen (as quoted in de Swart (1976a), p. 35) and Dragálin (1988), pp. 134–135; Gielen's construction is closest to van Dalen's. The (weaker) point that BKS and Church's Thesis are incompatible was first made in print by Myhill (1966, pp. 296–297), and taken up in the influential Troelstra (1969), p. 100.

[63]E.g., Troelstra and van Dalen (1988), p. 236, in particular: $A \leftrightarrow \exists n(\Box_n A)$. Intuitionistically, this is not difficult to justify; see the discussions of the topic in Dummett (2000), section 6.3, and van Atten (2004), chapter 5. (By the considerations in Sundholm and van Atten (2008), and also in Sundholm's contribution to the present volume, the principle cited should in fact be presented as a pair of (proof, not inference) rules, rather than as a bi-implication as understood in Natural Deduction. Note that the explanation usually given of the principle as cited is in effect that of the rules.)

[64]The equivalence would be best understood as an extensional one, so as to forestall paradoxes that might appear if one would straightforwardly render the sentential operator $\Box_n$ by a provability predicate. Alternatively, one could use BKS instead of CS to construct the function, as mentioned in footnote 62. I thank Albert Visser for raising this issue and for his 'Répondez!'.

[65]Letter from Kreisel to MvA, August 19, 2006.

7.2.5.5 Choice Sequences

While rejecting choice sequences from his own point of view, in his reflections on finitism in Hilbert's sense Gödel was led to conclude that choice sequences should be acceptable on that position. In a draft letter to Bernays of July 1969, he wrote:

> it now seems to me, after more careful consideration, that choice sequences are something concretely evident and therefore are finitary in Hilbert's sense, even if Hilbert himself was perhaps of another opinion. (Gödel 2003a, p. 269)[66]

and with that draft he included the text of a footnote for the revision of the Dialectica paper, in which he stated:

> Hilbert did not regard choice sequences (or recursive functions of them) as finitary, but this position may be challenged on the basis of Hilbert's own point of view. (Gödel 2003a, p. 270)[67]

In the letter he actually sent at the end of that month, he did not include the text for the footnote, and wrote 'Hilbert, I presume, didn't want to permit choice sequences? To me they seem to be quite concrete, but not to extend finitism in an essential way' (Gödel 2003a, p. 271).[68] However, in D70 (which Bernays would still see) and D72, he chose the slightly weaker formulation 'a closer approximation to Hilbert's finitism [than using the notion of accessibility] can be achieved by using the concept of free choice sequences' (Gödel 1972a, p. 272, note c).

7.2.5.6 1968: The Dialectica Interpretation as a Phenomenological Contribution to Intuitionism

Archive material shows that the foundation of the Dialectica Interpretation on a notion of 'reductive proof', well known from the publication of D72 in the *Collected Works*, was preceded by an attempt to construe the Dialectica Interpretation as a specifically intuitionistic result in the sense of Brouwer, and that both attempts were meant as applications of Husserl's phenomenology.[69]

Documentation of Gödel's phenomenological but not specifically intuitionistic approach to reductive proof is presented in Sect. 7.2.5.8 (p. 200).

---

[66] er scheint mir jetzt, nach reiflicher Überlegung, dass die Wahlfolgen etwas Anschauliches u[nd] daher im Hilbertschen Sinn Finites sind, wenn auch Hilbert selbst vielleicht anderer Meinung war. (Gödel 2003a, p. 268)

[67] This corresponds to item 040498 in 9b/148.

[68] Hilbert wollte Wahlfolgen wohl nicht zulassen? Mir scheinen sie durchaus anschaulich zu sein, aber den Finitismus nicht wesentlich zu erweitern. (Gödel 2003a, p. 270).

[69] These phenomenological projects were overlooked in the research for van Atten and Kennedy (2003), to which this part of the present paper should be considered an addendum.

The manuscript D68 contains various references to phenomenology. It is referred to as a possible method for developing a wider, yet no less convincing notion of constructivism than that of the formalists' (and, implicitly, of Gödel's former self of the 1930s). Having stated the intuitionistic conception, he adds this footnote:

> This explanation describes the standpoint taken, e.g., in A. Heyting's development of intuitionistic logic (see footn. 13).[70] Formalists in their consistency proofs are aiming at a stricter version of constructivism, which however has never been precisely defined. The most important additional requirement would no doubt be that the use of the term 'any' is restricted to totalities for which procedures for constructing all their elements are given. Also (which should be a consequence of this requirement) the conceptual selfreflexivities occurring not only in classical, but also in intuitionistic mathematics (e.g., that a numbertheoretic proof may contain the concept of numbertheoretic proof) are to be avoided. The hierarchy mentioned in footn. 6 is an example of this stricter constructivism, possibly even in case it is extended beyond ε_0, which can be done by treating as one step any sequence of steps which has been recognized as permissible (e.g., any ε_0 sequence of steps). The concepts of 'accessible', Brouwer's ordinals, and similarly defined classes would seem to need further analysis (perhaps in terms of the just mentioned hierarchy) in order to be strictly constructivistic.
> Not even the PFN functions[71] (if defined as below on p.   are strictly constructivistic (see p.   ). This makes one suspect that the aforementioned requirements of strict constructivism are too restrictive. Perhaps confining the extensions of concepts to sets that can somehow be 'overlooked' and avoiding selfreflexivities in the primitive terms are not the only means of reaching completely convincing proofs. Phenomenological clarification of the basic elements of our thinking should be another very different, and perhaps less restrictive, possibility. [references left open by Gödel; 9b/141, item 040450, p. 9F, 9.1F, 9.2F; 1968]

There is also some hesitation:

> As far as obtaining incontrovertible evidence [as the basis of a consistency proof of classical analysis] is concerned, what is needed would be phenomenological analysis of mathematical thinking. But that is a rather undeveloped field and there is no telling what future work in it may bring to light. [9b/141, item 040450, p. 12; 1968]

But later on in the same set of draft pages, in a passage here labelled (I), furthergoing claims are made:

> (I) On the other hand the interpretation of T used in this paper yields a consistency proof based on a new intuitionistic insight, namely the immediate evidence of the axioms and rules of inference of T for the computable functions defined above. Note that, as our analysis has shown, this insight is based on psychological (phenomenological) reflection, whose fruitfulness for the foundations of mathematics is thereby clearly demonstrated. [9b/141, item 040450, pp. 21–22; 1968]

---

[70] [In his footnote 13, Gödel refers to Heyting (1934, p.14).]

[71] [Primitive recursive functions of finite type over the natural numbers [9b/141, item 040450, p. 24; 1968]]

The following four (overlapping) topics evoked by or related to (I) will be commented on and illustrated by other passages:

I1. Psychological and logical reflection,
I2. psychology and intuitionism,
I3. psychology and phenomenology,
I4. the Dialectica Interpretation as an application of phenomenology,
I5. Gödel's 'analysis'.

I1. Psychological reflection is contrasted with logical reflection:

> We comprise both kinds of concepts (i.e., those obtained by logical and those obtained by psychological reflection under the term 'abstract', because the thoughts in question always contain abstract elements, either as their object or at least as being used. However, finer distinctions are of course possible. E.g., the concept of idealized finitary intuitions (see p.   above) evidently is formed by psychological reflection. [reference left open by Gödel; 9b/141, 040450, 1968, p. 6.1F]

The contrast is again described in the notes towards D70:

> Husserl Note that conc[erning] abstr[act] concepts[72] one has to distinguish thoughts & their content (obtained by psychol[ogical] & log[ical] reflection resp[ectively]) The former (to which int[uitionists] try to confine themselves[73]) are occurrences in the real world & therefore are in a sense just as concrete as [...][74] of symbols which should make them all the more acc[eptable] to finitists. [9b/148.5, item 040498.60]

Husserl discusses this distinction in, for example, in sections 41 and 88 of *Ideen I* (Husserl 1950), a work that Gödel owned and knew well; these are titled 'The really inherent composition of perception and its transcendent object' and 'Real and intentional components of mental processes. The noema', respectively.[75] It is the distinction between (mental) acts as concrete occurrences in time and their intended objects as such. The distinction applies to all thoughts, but Gödel's concern is with those that are in some sense abstract:

> We comprise both kinds of concepts (i.e. those obtained by logical and those obtained by psychological reflection) under the term 'abstract', because the thoughts in question always contain abstract elements, either as their object or at least as being used. However, finer distinctions are of course possible. E.g., the concept of idealized finitary intuition (see p.   above) evidently is formed by psychological reflection. [page reference left open by Gödel. 9b/141, item 040450; 1968]

---

[72][above 'concepts', Gödel wrote: 'entities']

[73][Also: 'speaking (as intuitionists [...] do) of thoughts as occurrences in spacetime reality (instead of their content) the objectivation (in the statements of the theory) of abstract entities and existential assertions about them are avoided and, moreover, the content of the thoughts to be admitted, although itself something abstract, always refers to something concrete, namely other thoughts or symbols or actions'. 9b/148.5, item between 040498.39 and 040498.43.]

[74]Unreadable word; 'comb[inations]'?

[75]'Der reelle Bestand der Wahrnehmung und ihr transzendentes Objekt' and 'Reelle und intentionale Erlebniskomponenten. Das Noema'. Translations taken from Husserl (1983); the second one is modified.

An example of a thought that is not directed at an abstract object but nevertheless uses an abstract element would be an insight about infinitely many concrete acts. As an infinity of acts cannot actually be carried out by us, they cannot all be concretely represented in a thought, and we have to represent them abstractly. According to Gödel, both finitary mathematics and intuitionistic mathematics arise from psychological reflection; the difference between them is that in the latter, the abstract elements that can be used in thoughts about the concrete acts themselves also become objects of the theory.

I2. Gödel emphasizes that the kind of psychology of which in (I) he considers intuitionism to be a form is not empirical psychology:

> Of course, in order to carry through this interpretation accurately and completely, a much more careful examination of the situation would be necessary. In particular the question would have to be answered why intuitionistic mathematics does not become an empirical science under this point of view. Rougly speaking, the answer is the same as that to a similar question about metamathematics as the science of handling physical symbols (although the situation is much more involved in our case). The relevant considerations in both cases are these: 1. There exist necessary propositions about concrete objects, e.g., that parts of parts are parts. 2. Mathematical propositions (in particular existential propositions) in this interpretation may be looked upon as implications whose hypotheses are certain (evidently possible) general empirical facts. 3. Instead of speaking of the occurrences (in reality) of mental acts or physical symbols one may speak of their individual forms (which determine their qualities in every relevant detail). In the special intuitionistic considerations given in the present paper the psychological interpretation has not been used throughout, e.g., we speak of rules (in the sense of procedures decided upon as to be followed) governing mental activity, not only of mental images of such rules. [9b/141, item 040450, page 9.2; 1968]

Hence he could say to Toledo, as we saw above, that intuitionism is 'essential a priori psychology'.

I3. It is clear that Gödel, when in (I) he sees intuitionism as 'psychological (phenomenological) reflection' which is at the same time 'a priori', he is speaking of what Husserl called 'phenomenological psychology'. Husserl wrote extensively about this in two places that Gödel knew well: the *Encyclopedia Britannica* article[76] and in the last part of the *Krisis* (Husserl 1954), titled 'The way into phenomenological transcendental philosophy from psychology'.[77] And, as it happened, the Husserliana volume with Husserl's 1925 lectures *Phänomenologische Psychologie*

[76]Gödel may have read this before 1962, the year the original German manuscript was reprinted in the Husserliana edition (Husserl 1962); there is a library slip (9c/22, item 050103) requesting the relevant volume (17: 'P to Planting of Trees') of the 14th edition of the *Britannica* of 1929. There are also some reading notes in the same folder. For a different connection between Gödel and the Britannica article, see van Atten and Kennedy (2003), section 6.1.

[77] Der Weg in die phänomenologische Transzendentalphilosophie von der Psychologie aus.

came out in 1968; but I don't know whether Gödel got to see that when working on the new introduction to the Dialectica paper in the first months of that year.[78]

Phenomenological psychology describes mental phenomena and unlike empirical psychology is not concerned with individual concrete facts but with invariant forms they instantiate and which delineate the range of possible concrete facts. In other words, it deals with the essence of our psychology.[79]

Note that Gödel's conception of the intuitionistic subject here as a subject in a psychological and hence mundane sense can be challenged, on grounds that are no doubt clearer in Brouwer's writings than in Heyting's. For an argument that the intuitionistic subject is better understood as a transcendental subject in Husserl's sense, see van Atten (2004, ch.6, 2010, pp. 66–68).

I4. Given the above, the Dialectica Interpretation is in D68 meant to be an application of phenomenology that moreover belongs to intuitionism, because it is based on an insight into mathematical procedures understood as acts carried out in thought over time (noeses), which are then, in acts of reflection, objectified as such to become objects of the theory.[80]

I5. Among the notes Gödel made in 1968 in preparation for D68 is the following:

> Foundations: it is really incredible, how all important philosophical and psychological problems are actualized in a rigorous treatment of my system T, and how many important distinctions become clear. For example: evocation of the image of a procedure and application of the procedure; image of a rule and rule (one sees how 'flimsy'[81] the former is, and how 'iron' the latter); results of the intermediate steps and the operations of the intermediate steps; operation in the sense of a mental act and of a mathematical object (briefly: rule, image of a rule, application of a rule, image[82] of the application of a rule);

---

[78] 1968 is the copyright year. That is not necessarily the year the book became available.

[79] The project of a non-empirical (e.g., 'a priori', 'rational' or 'transcendental') psychology has a long tradition (e.g., Wolff, Kant); for Gödel, Husserl's version will have been attractive because it is closely related to transcendental phenomenology, to which Husserl considered it to be propaedeutic.

[80] 'Strictly speaking the construction of intuitive mathematics in itself is an action and not a science; it only becomes a science ... in a mathematics of the second order, which consists of the mathematical consideration of mathematics or of the language of mathematics' (Brouwer 1975, p. 61 n.1). ('Eigenlijk is het gebouw der intuïtieve wiskunde zonder meer een daad, en geen wetenschap; een wetenschap ... wordt zij eerst in de wiskunde der tweede orde, die het wiskundig bekijken van de wiskunde of van de taal der wiskunde is.' (Brouwer 1907, p. 98 n.))

[81] [In English in the original.]

[82] [I translate 'Vorstellung' as 'image' here, because that is the term Gödel uses in these manuscripts when writing in English. Spiegelberg (1965), a work that Gödel owned (2nd ed.) and knew well, translates it as 'representation', and a popular alternative is 'presentation'. (NB Cairns' recommendation, published in 1973, for the broadest Husserlian sense is '(mental) objectivation' (Cairns 1973, p. 131).) I take it that Gödel's choice of 'image' is motivated by a wish to avoid special terminology as much as possible, so as to avoid making his philosophical remarks seem more dependent on a particular philosophy than they are. To Wang he said, 'I am cautious and

definitional procedure to obtain the functions of *T*, procedure to compute the individual functions of T[83,84];

However, although D68 itself does contain what could be considered to be preliminary remarks to a (phenomenological) analysis, e.g.,

> By 'procedure' we mean here 'mental' or 'thought' procedures, i.e., the steps are (intuitionistically meaningful) ideas or mental images formed by mental acts on the basis of the preceding steps according to the rule of the procedure. Also 'starting with' or 'terminating with' means: starting or terminating with a mental image of .... For practical reasons the writing down and 'reading' of symbols (used only for denoting well determined thoughts) are also to be admitted as steps of the procedures. Of course the rules of the procedures are supposed to be such that each step is (in a repeatable manner) uniquely determined by the previous steps. [9b/141, 040450, p. 17]

the body of a detailed analysis, in terms of the concepts and distinctions mentioned in the previous quotation, is not to be found in it. There is further archive material around D68 awaiting transcription; but I assume that, if it contained substantial further analysis, that would have been included in the longhand draft. Without such an analysis, it is not clear that there will be any advantage in shifting to a specifically intuitionistic (noetic) perspective (as, in contrast, there is, for Brouwerians, when demonstrating the bar theorem). This will undoubtedly have played a major role in Gödel's eventual dissatisfaction with D68.

This ends my discussion of I1–I5.

In spite of the above, perhaps one doubts Gödel's description in (I) of the reflection that led to the fundamental insight of the Dialectica Interpretation as 'phenomenological', on the ground that he had obtained that insight long before his

---

only make public the less controversial parts of my philosophy' (Wang 1996, p. 235). Similary, Wang remarks that 'Gödel's desire to shun conflict also affected his published work. He would make great efforts to present his ideas in such a form that people with different perspectives could all appreciate them (in different ways).' (Wang 1996, p. 235). (I thank Nuno Jerónimo for locating these comments.)]

[83] [Here the list stops, at the bottom of the left half of the page, and the right half of the page begins with a new remark.]

[84] 9b/148, item 040492. Transcription Eva-Maria Engelen; translation MvA. 'Gr[undlagen]: Es ist unglaublich, wie sämtliche wichtigen ph[ilosophischen] und psych[ologischen] Probleme bei genauer Behandlung meines Systems T aktualisiert [werden] und wie viele wichtige Distinct[ionen] klar werden: zum Beispiel: Evokation der Vorstellung eines Verfahrens und Anwendung des Verfahrens; Vorstellung einer Regel und Regel (man sieht wie 'flimsy' die erstere und wie 'ehern' die letztere ist); Resultate der Zwischenschritte und Operationen der Zwischenschritte; Operation im Sinn einer geistigen Handlung und eines mat[hematischen] Objekts (kurz: Regel, Vorstellung der Regel, Anwendung der Regel, Vorstellung der Anwendung der Regel); Def[initions-]Verfahren, um die Funktionen von T zu erhalten, Verfahren um die einzelnen Funktionen von *T* zu berechnen;'

turn to phenomenology around 1959.[85] But I don't think that Gödel here is making an implicit historical claim, but rather is using, on the occasion of a new presentation of his earlier insight, the framework that by then he had come to see as the best one for its philosophical reconstruction and explication. Gödel's philosophical remarks in the introduction to (both versions of) the Dialectica paper comfortably fit the description he had given of phenomenology and its use in his earlier text *1961?:

> Here clarification of meaning consists in focusing more sharply on the concepts concerned by directing our attention in a certain way, namely, onto our own acts in the use of these concepts, onto our powers in carrying out our acts, etc. [... It] is (or in any case should be) a procedure or technique that should produce in us a new state of consciousness in which we describe in detail the basic concepts we use in our thought. (Gödel 1961/?, p. 383)

However, having written (I), at some point Gödel put question marks next to it; recall his remark to Bernays of December 1968, quoted on p. 186 above, that he liked the 'new philosophical introduction' that he had written 'as little as the old'. There are various possibilities as to what type of doubt they express. For example, Gödel may have developed doubts whether this was what he could claim, or whether he would be able sufficiently to develop this claim in writing so as to be convincing to others, or whether this claim would be well received given the Zeitgeist as he perceived it,[87] or whether to convince others of his consistency proof it was even necessary to make and develop this specific claim. Finally, he may have developed second thoughts about presenting this work as an intuitionistic insight.[88]

If this hesitation is indeed a mark of the same discontent that Gödel expressed in his letter to Bernays of December 1968, then it is no surprise that the next year, he drafted the following (unsent) reply to an inquiry from van Dalen:

> My relationship with Intuitionism consists primarily in some theorems I proved about certain parts of intuitionistic mathematics in particular that published in Dial[ectica] 12.

---

[85]As Feferman points out in his introduction to the Gödel-Bernays correspondence, it is noteworthy that in this philosophically rich exchange, Husserl is never discussed (Gödel 2003a, p. 66, n. ax). (Gödel mentions phenomenology once, on August 11, 1961 (Gödel 2003a, p. 193).) Bernays will have known of Gödel's enthusiasm for Husserl early on: Bernays was in Princeton from November 1959 to April 1960, and came back for shorter visits around Easter 1961, in May 1961, and in the Spring of 1965.[86] A short text presented in 1963—in the middle of this period of visits—and published the next year, 'Begriffe des Phänomenologischen und das Programm der phänomenologischen Philosophie', shows Bernays rather critical of Husserlian phenomenology, in particular of époché, the possibility to see essences, and its foundational character. In his letters to Gödel, Bernays never mentions this text or the objections formulated in it; one possible explanation is that they had dealt with the topic in their conversations.

[86]See letters 26–31; 34–36; 38–39; and 52–53, respectively.

[87]That seems to have been the kind of reason why, in the published version of the supplement to the second edition of his Cantor paper, Gödel left out a hopeful reference to phenomenology that is present in the draft (8c/101, item 040311), while at the same time recommending Husserl to logicians in conversation. See van Atten and Kennedy (2003), p. 466.

[88]Elsewhere, I have argued that Gödel's program to employ transcendental phenomenology to found *classical* mathematics is misguided (van Atten 2010). But, by the positive argument in the same paper, his attempt to use it to enrich intuitionism, whether eventually successful or not, makes perfectly good sense.

> The question as to whether this paper is important for the foundations of Intuitionism I must leave for Intuitionists to answer. I did not write the paper from this point of view and some supplementation would be necessary in order to clarify it's [sic] relevance for the foundations of Intuitionism. [01/199, item 012891; underlining Gödel's]

This repeats of course Gödel's statement towards the end of the 1958 publication,

> Selbstverständlich wird nicht behauptet, dass die Definitionen 1–6 den Sinn der von Brouwer und Heyting eingeführten logischen Partikel wiedergeben. Wieweit sie diese ersetzen können, bedarf einer näheren Untersuchung. ((Gödel 1958, p. 286); translation below)

and moreover remains silent about his recently aborted attempt to develop the Dialectica interpretation as a specifically intuitionistic insight. Naturally, then, in D70 and D72 Gödel left the content of the 1958 statement unaltered:

> Of course it is not claimed that Definitions 1–6 express the meaning of the logical particles introduced by Brouwer and Heyting. The question to what extent they can replace them requires closer investigation. [D70, 9b/142, item 040451, p. 7; Gödel (1990, p. 280)]

Although Gödel thus abandoned the effort in D68 to give Dialectica a specifically intuitionistic content, he did not abandon his phenomenological approach to the philosophical deepening of that work; see Reductive Proof: Phenomenology and Demonstrations (p. 200).

7.2.5.7 Demonstrability and Impredicativity

In D70, Gödel included in his definition of 'computable function of type t_k' that it is 'intuitionistically demonstrable' that it is always performable; in the publication of 1958, this had been 'constructively recognizable'[89] (Gödel 1990, p. 245). Bernays read the galleys and on July 12, 1970 remarked to Gödel about this definition:

> Here the reader could well be taken aback, since your procedure is surely intended to avoid the concept of intuitionistic proof. It seems to me, however, that in fact you do not need that concept here at all, and that onloy a suitable reformulation is needed in order to make that clear. (Gödel 2003a, p. 281)[90]

In later letters, Bernays proposed two alternatives:

1. Replace 'is intuitionistically demonstrable' by 'follows directly from the definition of the function in question and those of the functions in the k-tuple' (Gödel 2003a, p. 286/287, October 12, 1970)

---

[89] konstruktiv erkennbare (Gödel 1958, p. 282).

[90] Hier könnte wohl der Leser stutzen, da doch Ihr Verfahren bezweckt, den Begriff des intuitionistischen Beweises zu vermeiden. Es scheint mir jedoch, dass Sie de facto hier diesen Begriff auch gar nicht brauchen un dass es nur einer geeigneten Umformulierung bedarf, um dieses zum Ausdruck zu bringen. (Gödel 2003a, p. 201)

2. Add a footnote saying that what is meant here is only that 'for the determination in question the methods of proof excluded by intuitionism must not be used' (Gödel 2003a, pp. 292–295, March 16, 1972)

Curiously, Gödel in D68 had foreseen Bernays' objection, answered it as he would later do in D72, and then rejected that answer:

> For the special application of the concept of computable function to be made in the present paper it is preferable to replace in its definition the term 'demonstrable' by 'evident on the basis of the definition of the procedure and previous definitions used in it'.
> [...]
> It might be objected that the concept of proof is used also in our interpretation, since 'demonstrability' occurs explicitly in the definition of CFI function.[91] | The answer is that, in constructing a model of *T*, 'demonstrable', in the definition of CFI function, may be replaced by 'evident without proof on the basis of the structure of the definitions.' [9b/141, 040450, pp. 19 and 23–24.]

But Gödel then crossed out the first passage, and wrote 'Wrong' next to the second.[92] I do not know what reason for doing so he had in mind; but it was evidently what motivated him to write 'intuitionistically demonstrable' in D70, which Bernays then got to see—unlike D68. By the time of D72, Gödel had dropped whatever his earlier objection had been, and replaced 'intuitionistically demonstrable' by 'constructively evident'.[93] It is not clear to what extent that was under the influence of Bernays' remarks.

In D68 Gödel already realized that there is an impredicativity in his system *T*:

> There are functions of lower type which (within *T*) can only be defined by using functions of much higher types. This 'impredicativity' is perfectly legitimate, also from the constructivist point of view. It was in substance admitted even in Princ[ipia] Math[ematica] 2nd ed. p.   . And indeed the fact that the concept of functions of high type is defined in terms of that of functions of low type in no way precludes an inverse relationship for individual functions; i.e., the chain of definitions of a PCN function may go up and down in the system of types. What can be concluded from this state of affairs is only that the PCN functions (if introduced as above) are not strictly constructivistic in the sense of footn. [references left open by Gödel. 9b/141, item 040450, pp. 25–26; 1968]

Note that Gödel here harks back to his Russell paper[94]:

> In the second edition of *Principia*, however, it is stated in the Introduction (pages xl and xli) that 'in a limited sense' also functions of a higher order than the predicate itself (therefore also functions defined in terms of the predicate as, e.g., in *p* 'κ ε κ) can appear as arguments of a predicate of functions; and in Appendix B such things occur constantly. This means that

[91]['CFI' is Gödel's abbreviation in D68 for 'computable functional of finite type' [9b/141, item 040450, pp. 23—24; 1968]]

[92]'Falsch', in shorthand.

[93]In note k of D72 (Gödel 1990, p. 275, note h), however, he quoted that part of the definition as 'constructively evident or demonstrable'; I assume this was left in inadvertently.

[94]Gödel studied his Russell paper when working on the Dialectica paper. This is clear from the remarks on the loose sheet inserted with one of the four offprints that Gödel owned, offprint D in the 'Textual notes' in the *Collected Works*, Gödel (1990, p. 315–322); the remarks in question are on p. 320 and p. 321. NB Correction: The note on the title page of D does not say 'gelesen bis p. 135 oben' (Gödel 1990, p. 320), but 'gelesen bis p. 138 oben'.

> the vicious circle principle for propositional functions is virtually dropped. This change is connected with the new axiom that functions can occur in propositions only 'through their values', i.e., extensionally, which has the consequence that any propositional function can take as an argument any function of appropriate type, whose extension is defined (no matter what order of quantifiers is used in the definition of this extension). There is no doubt that these things are quite unobjectionable even from the constructive standpoint (see page 136), provided that quantifiers are always restricted to definite orders. (Gödel 1944, p. 134)

The sense in which in D68 the PCN functions are said not to be strictly constructivistic, then, is the sense in which it would be demanded that functions are generated from below exclusively.

In D70, with the introduction of the notion of reductive proof, it was clear to Gödel that notion would not serve to avoid this impredicativity, and the corresponding remark is phrased thus:

> In particular, there exist functions of lower type which, within *T*, can only be defined in terms of functions of higher types. This is a kind of impredicativity. True, it is only one of those weak impredicativities that are admitted even in Princ[ipia] Math[ematica] 2nd ed. p. [XL],[95] ff. In our proofs of the axioms of *T* this impredicativity appears in the fact that the concept of reductive proof may itself occur in reductive proofs (just as in Heyting's logic the general concept of proof may occur in a proof).

It is the impredicativity mentioned in the final sentence that Gödel refers to as the 'unavoidable self reflexivities'[96] in his letter to Bernays of December 22, 1970 (Gödel 2003a, p. 290/291). From Gödel's letter to Bernays of 2 years later (December 26, 1972), it is clear that he remained convinced that, all the same, the notion of reductive proof was an epistemic advance over Heyting's Proof Interpretation:

> I also thank you very much for your letter about the question whether the general intuitionistic concept of proof is necessary for the intuitionistic interpretation of my system *T* (which would make my interpretation of the logical operators epistemologically worthless). I think that that is *not* the case, but rather that a *much* narrower and in principle *decidable* concept of proof suffices, which I introduced in note k[97] of the translation of my *Dialectica* paper and called 'reductive provability'. But to carry that through satisfactorily in detail is not all that easy, mainly on account of the *non-eliminable* impredicativity also of this narrower concept of proof, which is closely connected with the impredicativity of the concept of function that you mentioned. It is doubtful whether carrying it through would be worth the trouble. Up to now, therefore, I have not been able to make up my mind to do it, although the further pursuit of that question could perhaps contribute in an essential way to the clarification of the foundations of intuitionism. (Gödel 2003a, p. 301)[98]

---

[95] [The typescript erroneously has 'XI', as had the original publication of Gödel's Russell paper (Gödel 1944, p. 134).]

[96] [[die] unvermeidlichen "self reflexivities".]

[97] [Presented, in the later version of D72, as note h in Gödel (1972a).]

[98] Ich danke Ihnen auch bestens für Ihren Brief über die Frage, ob der allgemeine intuition[istischen] Beweisbegriff für die intuition[istischen] Interpretation meines Systems *T* nötig ist (was meine Interpretationder logischen Operatoren erkenntnistheoretisch wertlos machen würde). Ich glaube, dass das *nicht* der Fall ist, sondern dass ein *viel* engerer

An as yet unpublished assessment by Gödel of this situation, dating from 1974, will be presented in Sect. 7.2.5.9.

7.2.5.8 Reductive Proof: Phenomenology and Demonstrations

Although Gödel abandoned the attempt in D68 to construe the Dialectica Interpretation as intuitionistic in the noetic sense, the shift to the notion of reductive proof employed in D70 and D72 still depended on phenomenology, and still marked a rapprochement to Brouwerian intuitionism.

That Gödel could continue to use the phenomenological method is not surprising: as a study of consciousness, phenomenology is, at least at a certain level of generality, compatible with different views as to what mathematics is and how it is related to consciousness.[99] This is clear from the following group of notes[100]:

4. In my interpretation there are [...] no such iterations of implications, as little as of universal propositions, because the premisse always contains Red[uctive].[101]

noema /

5. Does in the intentional object of the knowledge of such an imp[lication] [...][102] $\supset$ $c = g(d)$ the concept of insight occur? No, because the noema of the cognitional act contains both of these in a certain relation.

[...]

7. A good example of the distinction between (and transition from) noema to noesis [is] the int[uitionistic] (Heyt[ingian]) interpretation of imp[lication].

(u[nd] im Prinzip) entscheidbarer Beweisbegriff genügt, den ik in Note k der Übersetzung meiner Dialectica [arbeit] eingeführt u[nd] 'reuktive Beweisbarkeit' genannt habe. Aber das im einzelnen befriedigend durchzuführen, ist nicht ganz leicht, hauptsächlich wegen der *nicht eliminierbaren* Imprädikativität auch dieses engeren Beweisbegriffes, welch miet der von Ihnen erwähnten Imprädikativität des Funktionsbegriffes nahe zusammenhängt. Est ist zweifelhaft, ob die Durchführung die Mühe lohnen würde. Ich habe mich daher bis jetzt nicht dazu entschliessen können, obwohl die weitere Verfolgung dieser Fragen vielleicht wesentlich zur Aufklärung der Grundlagen des Intuitionismus beitragen könnte. (Gödel 2003a, p. 300)

[99]I write 'at a certain level of generality', because this compatibility may or may not be preserved when making one's conception of phenomenology more specific. In van Atten (2010) I argue that, in particular, if one's conception of phenomenology is that of the transcendental Husserl (of, roughly, the 1920s and 1930s), then intuitionistic mathematics is compatible with, and moreover part of, phenomenology, whereas classical mathematics is neither.

[100]In the archive, these are not kept with the drafts and galleys for the revised Dialectica paper, but in a folder named 'Dialectica interpretation' in the section 'Other loose manuscript notes'.

[101][There are notes in which Gödel writes antecedents as '$Red(p)$', for 'p is reductively provable' (which for given p is decidable, or should be once the notion of reductive proof has been sufficiently clarified).]

[102][The antecedent is almost unreadable, but it seems safe to say that Gödel here gives an example of an implication in T.]

[...]

12. Important:

    [...]

    2. All propositions must be accepted as meaningful objects[103] (= int[entional] Obj[ects] = Noemata) and likewise [the] chains of evidences for them, where the preceding propositions have indeed been seen to be true before, hence no assumption.
    3. The chains of evidences, which appear when expl[icating] Red[uctive] Proof, are of this kind. That is also exactly as for finitary proofs (this is so for the theorems and inferences in P[104] in general)[105]

[103]Ground objects and (x) is therefore an operation binding ground variables that leads to objects.

[104][P is the system that was going to be named T' in D72 (see the manuscript for D70, 9b/142, item 040452, insertion to note k3. In the circulated typescript, this is the footnote on p. 13).]

[105]11b/6, item 060039. Transcription Eva-Maria Engelen, Robin Rollinger, and MvA. Translation MvA.

4. In meiner Interpretation gibt es [...] keine solchen Iterationen von Impl[ikationen], ebensowenig wie die von Allsätzen, weil in der Prämisse immer Red[uctive] steht.

                Noema /

5. Kommt im intentionalen Objekt der Erkenntnis einer solchen Imp[likation] [...] $\supset c = g(d)$ der Begriff der Einsicht vor? Nein, | denn der Akt der Erkenntnis hat diese beide im Noema in einem [be]stimmten Zusammenhang.

[...]

7. Ein gutes Beispiel für Unterscheidung von (und Übergang von) Noema zu Noesis [ist] die int[uitionistische] (Heyt[ingsche]) Interpretation der Imp[likation].

[...]

12. Wichtig:

    [...]

    2. Alle Sätze müssen als sinnvolle Objekte[106] (= int[entionale] Obj[ekte] = Noemata) anerkannt werden und ebenso [die] Evidenzketten für solche wo die vorausgehenden Sätze vorher tatsächlich eingesehen sind, also keine Annahme.
    3. Die Evidenzketten, welche bei der Expl[ikation] von Red[uctive] Proof herauskommen, sind von dieser Art. Das ist also ganz genau so wie bei finiten Beweisen (wie überhaupt die Sätze und Schlüsse in P)

[106]Grundobjekte und (x) ist also eine Grundvariablen bindende Operation, welche zu Objekte[n] führt.

When working the ideas into a revision of D70's note k (not included in D72), this became:

> As for item 2. ['the meaning of the implications of the form "If $x, y, \ldots$ have certain types, then ..." occurring implicitly both in the definition of 'computable of type t' and in the axioms and theorems of T'] it is first to be noted that implication occurs only in this form: 'If the procedure A yields the result a, then the procedure B yields the result b' where it need not be known whether procedures A or B yield any result at all, even though they are supposed to be defined with perfect precision.[107] For, also the statement '$(x)\phi(x)$ is reductively provable' (which is the only way in which quantification occurs in the interior of formulas) means that a certain procedure of checking the chain of defi|nitions of the concepts in ϕ yields a certain result. But such implications can be interpreted to mean: 'If I (the reasoning mathematician) carried out the procedure A and obtained the result a then, if I carry out the procedure B, I shall obtain the result b', where the 'ifs' here mean a truthvalue function, i.e., 'either the implicans is false or the implicatum true'. This entails that, in the last analysis, the implication in question means that a certain procedure involving both A and B yields a certain result, whenever carried out.' [9b/145, item 040458, pp. 2–3]

In this draft towards a publication, the phenomenological terminology used in Gödel's private notes has disappeared. This seems to me to be intentional; see also footnote 82 above.

Note how Gödel's explanation of implication (with respect to reductive proof), is given by demonstration-conditions, that is, conditions in terms of procedures that, actually or hypothetically, have been carried out; as opposed to proof-(object-)conditions, which are given in terms of properties of proofs independently of the actual or hypothetical fact that we know this proof-object (Sundholm 2007).

In Sundholm and van Atten (2008) arguments are given why demonstration-conditions, not proof-conditions, are required for a correct reading of Brouwer. On the one hand, then, the Dialectica Interpretation on the basis of reductive proof is much closer to Brouwerian intuitionism than to alternative constructive foundations (see Sect. 7.2.5.9, remark on [A], p. 205). On the other hand, the combined effect of items 5 and 7 is to distance the notion of reductive proof from that of intuitionistic proof in the specifically noetic sense of Brouwer and Heyting (see item I4 above, p. 194). As a consequence, Brouwer's specific mentalism about mathematics as a whole makes certain types of argument available to him that Gödel cannot use.[108]

7.2.5.9 Gödel's Two 1974 Assessments of the Dialectica Interpretation

The *Collected Works* include a draft letter from Gödel to Frederick Sawyer, already mentioned on p. 183, which was probably written not long after February 1, 1974. Gödel there claims that (because of the employment of the notion of reductive proof),

[107] [The part from 'where' to the end is a later insertion: item 040462, k(2) +.]

[108] See Sundholm and van Atten (2008, section 6), and the remark on item [C] on p. 204 below.

the implicit use of 'implication' and 'demonstrability' occurring (through the words 'immer ausfuehrbare'[109] and 'erkennbare' in the definition of 'computable function of finite type' on p. 282–283[110] does *not* give rise to any circularity. (Gödel 2003b, p. 211, original emphasis)

However, the archives also contain the following note, dated February 11, 1974, not long after the day on which Gödel must have received Sawyer's letter (the labels [A], [B], [C] are mine):

February 11, 1974

[A] My Dialectica paper with the notion of reductive proof does not give an interpretation that excludes the paradoxes (hence the foundation not essentially better than Heyting, namely for this reason, that for example the general concept of computable number-theoretic function occurs and this speaks of a chain of definitions (hence the definition $x \in a \equiv \sim x \in x$ may occur). The difference is only that the concept of evidence is applied only to the correctness of a definitions, not to the correctness of a proof. That is to say, they do not exclude the 'vastness' of the domain in question, as the concepts 'number-theoretic evidence', 'type-theoretical evidence', 'evidence with respect to functional of finite type' etc. do.[111] These concepts as primitive concepts are 'vague'. But perhaps the admissible propositions can be defined precisely (and these would then be a constructed set like the natural numbers, Gentzen), but the concept of number-theoretic meaningful proposition would presuppose the concept of number-theoretic meaningful proof, as it may contain B, hence [is] circular. What is thus accomplished, is threefold:
[B]

1. 'correct proof' replaced by 'correct Def[inition]',
2. the proof is mathematically more direct (many 'convolutions' are avoided),
3. the problem of being and having for existential propositions is solved.

[C] Some normal form theorem for proofs might follow (from 3.), from which bar induction might follow?? The impossibility to prove something that is absolutely unprovable might follow from an idealisation of proofs using certain primitive notions and then one could define proofs as mathematical proofs by these means and that would suffice for the consistency proof.

But all this, to make sense, presupposes that one has resolved the paradox $\sim x \in x$.[112]

[109]I.e., *if* the arguments are computable. [original emphasis]

[110] Wenn die Begriffe "berechenbare Funktion vom Typus t_0", "berechenbare Funktion vom Typus t_1", ..., "berechenbare Funktion vom Typus t_k" (wobei $k \geq 1$) bereits definiert sind, so wird eine berechenbare Funktion vom Typus $(t_0, t_1, \ldots, t_k)$ definiert als eine immer ausführbare (und als solche konstruktiv erkennbare) Operation, welche jedem k-tupel berechenbarer Funktionen der Typen $t_0, t_1, \ldots, t_k$ eine berechenbare Funktion vom Typus t_0 zuordnet. Dieser Begriff ist als unmittelbar verständlich zu betrachten, vorausgesetzt dass man die Begriffe "berechenbare Funktion vom Typus t_i" ($i = 0, 1, \ldots, k$) bereits verstanden hat.

[111][Because in the first two cases the proofs are generated from below, and in the third case (the notion used in the main text of Gödel 1958, 1972a) the evidence is taken to be immediate.]

[112]10a/40, item 050136. Transcription Eva-Maria Engelen; translation MvA. The bars and underlining are Gödel's.

I wish to make the following comments on the parts of this note, starting with [B] and [C].

Item 3 in [B] is a reference to Leibniz' theory of truth, in whom Gödel found his inspiration for the notion of reductive proof; I refer to van Atten (forthcoming) for the argument for this claim, with documentation from the archive. Gödel's adaptation of Leibniz' idea of reductive analysis is indicative of his commitment to Leibnizian ideas even at that late stage of his career.[113]

In [C], 'Bar Induction' is mentioned with an eye on Spector's consistency proof of analysis (Spector 1962), which uses a (generalized) form of Brouwer's principle.[114] The strategy that Gödel proposes here is to find a canonical form for

---

11. II. 74

[A] Meine Dial. Arbeit mit dem Begriff des reduktiven Beweis[es] gibt keine die Parad[oxien] ausschließende Interpretation (daher die Fundierung nicht wesentlich besser als Heyting und zwar deswegen, weil zum Beispiel der allgemeine Begriff der berechenbaren zahlentheoretischen Funktion vorkommt und dieser von irgendeiner Def[initions]-Kette spricht (also die Def[inition] $x \in a \equiv \sim x \in x$ kann vorkommen). Der Unterschied ist nur, dass der Begriff Evidenz nur auf Richtigkeit einer Def[inition] nicht auf Richtigkeit eines Beweises angewendet wird. Das heißt also, sie schließen nicht die 'vastness' des betracht[eten] Bereichs aus wie das Begriffe 'zahlentheoretische Evidenz', 'typentheoretische Evidenz', 'Evidenz hinsichtlich Funktion endlichen Typs' etc. tun. Diese Begriffe als Grundbegriffe sind 'vage'. Aber vielleicht kann man präzise die erlaubten Sätze definieren (und diese wären dann eine konstruierte Menge wie die natürlich[en] Zahl[en], Gentzen), aber der Begriff des zahlentheoretisch sinnvollen Satzes würde | den Begriff des zahlentheoretisch sinnvollen Beweis[es] voraussetzen, da er B enthalten kann, also zirkulär [ist]. Was also geleistet wird, ist dreierlei:
[B]

1. 'richtiger Beweis' ersetzt durch 'richtige Def[inition]',
2. der Beweis ist mat[hematisch] direkter (es werden viele 'Verschlingungen' vermieden,
3. das Probl[em] von Sein und Haben für Ex[istenz]sätze wird gelöst.

[C] Es könnte da irgendein Normalform-Th[eorem] für Beweise folgen (aus 3.), aus welchem der Bar Ind[uktion] folgen könnte ?? Die Unmöglichkeit eine absolute Unbeweisbarkeit zu beweisen, könnte folgen aus einer Idealisierung der Beweise mit gewissen Grundbegriffen und dann könnte man Beweise definieren als mat[hematische] Beweise mit diesen Mittelnund das würde genügen für den Widerspruchsfreiheitsbeweis.

Aber all das, damit es Sinn hat, setzt voraus, dass man der Parad[oxie] $\sim x \in x$ aufgelöst hat.

[113]Given this influence, it would be interesting also to look at Gödel's notion of reductive proof in relation to his remarks on analyticity of mathematics and Leibniz at the end of his Russell paper of 1944 (quoted on p. 184 above), and to the brief exchange on this in the Gödel-Bernays correspondence (Gödel (2003a, pp. 194, 200), and also p. 57 of the introduction); but I will not do this here.

[114]As is clear from the subject headings in Gödel's mathematical *Arbeitshefte*, conveniently listed in Dawson and Dawson (2005), Gödel closely studied Brouwer's interpretation of analysis. At the time of writing this, Jan von Plato has announced a talk (at the conference in Aix-en-Provence

proofs of the antecedent in the principle, like Brouwer; but unlike Brouwer, for Gödel this canonical form cannot be defined in noetic terms, as discussed above.

[A] As documented above Sect. 7.2.5.7, Gödel knew in D68 that T was impredicative, and also, in D70, that his new notion of reductive proof could not remove this. Gödel was also well aware of alternative foundations of intuitionistic logic and arithmetic that had been proposed from the late 1960s on, in which impredicativity was avoided.

In particular, Gödel of course knew the work by Kreisel and by Goodman on the Theory of Constructions,[115] which however never led to a satisfactory development; Kreisel's version was inconsistent, and in Goodman's version a proof of $A \rightarrow B$ is no longer a construction that is applicable to *any* proof of A. One should also mention here[116] the theory of constructions developed in response to these problems in Scott's 'Constructive validity' (1970). When it ran into problems over decidability, Gödel and Kreisel insisted that one accept abstract proofs and have a 'proof predicate' as a decidable propositional function over the universe of all of them; see the postscript to Scott's paper.[117]

By 1974, Gödel had also studied Howard's seminal manuscript of 1969 (later published as Howard 1980) on what has become known as the Curry-Howard isomorphism.[118] But Gödel wanted to accept abstract proofs as objects in the theory,

---

in July 2013) on these notes in relation to Gödel's thoughts about Gentzen's work, equally documented in these notebooks, in particular with an eye on the question to what extent Gödel may have anticipated Spector's result. I will therefore not attempt to say more about the matter here.

[115]There are reading notes on Kreisel (1965) in 11c/28, item 060369, and on Goodman (1970) in 10a/40, item 050142.

[116]As Sundholm urged me to do.

[117]It is remarkable that Kreisel in his long paper on Gödel and intuitionism (Kreisel 1987b) refers to neither his own, nor Goodman's, nor Scott's work on the theory of constructions.

[118]See footnote 15 on p. 173 above. In the same email referred to there, Prof. Howard also recalls the following conversation with Gödel, probably in their first meeting during that sabbatical:

> Gödel: 'You should extend your theory of constructions to transfinite types in such a way as to get a functional interpretation of set theory (ZFC).'
> Me: 'I made such an attempt a couple of years ago and concluded that, to carry this out, I would have to learn more set theory.'
> Gödel: 'So, do it.'
> Me: 'Learning a sufficient amount of set theory appears to be a daunting task. There are a lot of papers.'
> Gödel: 'Very little of a *substantial* nature has been done. In fact, if you just read my two papers, that may be sufficient.'
> At that point, he got up, walked across the room to a filing cabinet, pulled out reprints of the two papers (Proc. Nat. Acad. Sciences 1938, 1939) [Gödel (1938a, 1939)] and handed them to me, saying, 'Here is what you should read. You may keep these.' [Story 5, p. 83 of Howard, unpublished]

Prof. Howard comments (in the same email):

> Presumably what he had in mind in his first remark was that if my little theory of constructions is extended to transfinite types, in a natural way, as far into the transfinite

and, as Artemov (2001, p. 4) observes, 'as *proof objects* Curry-Howard λ-terms denote nothing but derivations in Int [i.e., formalized intuitionistic propositional logic] itself and thus yield a circular provability semantics for the latter'.

In the notion of reductive proof, Gödel had found, he believed, the right notion that is decidable and narrower than the general notion of intuitionistic proof, with the three advantages listed in [B]. As mentioned above (p. 202), this depends on understanding 'proof' as demonstration (i.e., acts that have been carried out) instead of (knowable but perhaps unknown) proof-objects. This marks a fundamental difference with the constructive foundations mentioned above and those inspired by them (notably Martin-Löf's Constructive Type Theory).

But the formulation of this view evidently did not lead Gödel finally to publish the revised paper. That is not surprising; already in December 1970 he had written to Bernays that 'The time of publication seems to me to be less important than the improvements to the text', and he certainly didn't have the full details this time either.[119] Also, his bad health at the time may have prevented him from doing substantial further work in any case.

Acknowledgements This is the revised and much extended text of the talk with the same title given at the conference 'Calculability and constructivity: historical and philosophical aspects' of the International Union of the History and Philosophy of Science (Joint Session of the Division of Logic, Methodology and Philosophy of Science and of the Division of the History of Science and Technology), Paris, November 18, 2006. Much of that talk was derived from a manuscript that has in the meantime appeared as part of the present author's contribution to van Atten and Kennedy (2009) (written in 2005). Other versions of that talk were presented at the plenary discussion 'Gödel's Legacy' at the ASL European Summer Meeting in Nijmegen, August 2, 2006 and at seminars in Nancy (2005), Tokyo (2006), Utrecht (2006), and Aix-en-Provence (2007). I am grateful to the respective organisers for the invitations, and to the audiences for their questions, criticisms, and comments.

The quotations from Gödel's notebooks and lecture notes appear courtesy of the Kurt Gödel Papers, The Shelby White and Leon Levy Archives Center, Institute for Advanced Study, Princeton, NJ, USA, on deposit at Princeton University. I am grateful to Marcia Tucker, Christine Di Bella, and Erica Mosner of the Historical Studies-Social Science Library at the IAS for their assistance in finding anwers to various questions around this material. In the study of Gödel's

> as possible, the resulting theory would provide an interpretation of a part of ZFC (or a constructive version of a part of ZFC) which would be significally weaker than ZFC itself. Hence one would have shown an essential limitation on what could be achieved by Brouwer's ideas. In other words, do to Brouwer's program what Gödel had done to Hilbert's program. At least, that was my impression at the time. I seem to recall that he actually said something to that effect, but I don't have any quotation, in my notes for Amy [as part of the preparation for the article Shell-Gellasch (2003)], of him saying that to me.

Probably in the Spring of 1973, Gödel encouraged Howard to read Girard's 1972 thesis to get some ideas towards such an extension to transfinite types. (Details in story 16, p. 110 of Howard, unpublished.) Aczel's interpretation of CZF in Martin-Löf's Constructive Type Theory (Aczel 1978) may be seen as an execution of this project.

[119] Der Zeitpunkt des Erscheinens scheint mir weniger wichtig zu sein als die Textverbesserungen (Gödel 2003a, p. 290/291).

notes in Gabelsberger shorthand, I have been able to consult Cheryl Dawson's transcriptions, which she generously made available to me; these were also useful to Robin Rollinger and Eva-Maria Engelen, to whom I am greatly indebted for additional, speedy help with the shorthand, also concerning previously untranscribed passages. Access to the microfilm edition of the Kurt Gödel Papers was kindly provided to Rollinger, Engelen and me by Gabriella Crocco. The present paper is realized as part of her project 'Kurt Gödel philosophe: de la logique à la cosmologie', funded by the Agence Nationale de Recherche (project number BLAN-NT09-436673), whose support is gratefully acknowledged.

Gödel's letters to his brother quoted here are part of a collection of letters that was found in 2006. I am grateful to Matthias Baaz and Karl Sigmund for bringing this correspondence to my attention, and for providing me with photocopies. These letters have been deposited at the Wienbibliothek im Rathaus, Vienna. The quotations appear courtesy of the Kurt Gödel Papers, The Shelby White and Leon Levy Archives Center, Institute for Advanced Study, Princeton, NJ, USA.

I am grateful to Dirk van Dalen, Georg Kreisel, Albert Visser, and, in particular, William Howard and Göran Sundholm, for comments, references, criticisms and discussion. An anonymous referee wrote a helpful report on an earlier version, and Jaime Gaspar helpfully sent a list of typing errors.

Prof. Howard kindly granted permission to quote from the reminiscences he sent me; some of these come from the notes he prepared for Amy Shell-Gellasch, who used them for her article Shell-Gellasch (2003). Those notes are now held at the Archives of American Mathematics, Dolph Briscoe Center for American History, University of Texas at Austin, as part of the William Howard Oral History Collection, 1973, 1990–2003. These Archives hold the copyright; quotations are by permission. I thank its staff member Carol Mead for her help and advice concerning this material and its use.

Appendix: Finitary Mathematics and Autonomous Transfinite Progressions

Naturally, the draft notes for the revision of the Dialectica paper also contain remarks that are not concerned with intuitionism as such, but with finitary mathematics.

> In support of the admission of abstract objects, note also that it is altogether illusory to try to eliminate abstractions completely, whatever the science in question may be. Even finitism in its strictest form does contain them, since every general concept is an abstract entity (although not necessarily an abstract concept, which term is reserved for concepts referring to something abstract). The difference between finitism and the envisaged extension of it only is that in the former abstractions occurring are only used, but are not made objects of the theory. So the question is not whether abstractions should be admitted, but only which ones and in what sense. It seems reasonable, at any rate, to admit as object of the investigation anything which is admitted for use. This leads to something like the hierarchy described in footn[ote] 7. [9b/148.5, item 040498.31]

The example referred to at the end is that of autonomous transfinite progressions, which Gödel describes in footnote 2 on p. 281 of the 1958 version and footnotes 4 and f of the 1972 version. On both occasions he refers to the formal work that appeared in print in Kreisel (1960, 1965); but in D68, he moreover writes that he had arrived at this idea when writing his incompleteness paper Gödel (1931), and had considered it finitary:

> That in Mon[ats]H[efte für] Math[ematik und] Phys[ik] 38 (1931), p. 197 I said that finitary mathematics conceivably may not be contained even in formalized set theory is due to the fact that, contrary to Hilbert's conception, I considered systems obtained by reflection on finitary systems to be themselves finitary. [9b/141, item 040450, p. 4F; 1968]

and, in a different version with the title 'Kreisel's hierarchy',

> How far in the series of ordinals this sequence of systems reaches is unknown. Evidently it is impossible to give a constructive definition and proof for its precise limit, since this ordinal would then itself be an admissible sequence of steps. When in Mon[ats]H[efte für] Math[ematik und] Phys[ik] 38 (1931) p. 197 I was speaking of 'conceivably' very powerful finitary reasoning, I was really thinking of this hierarchy, overlooking the fact that from a certain point on (and, in fact, already for rather small ordinals) abstract concepts are indispensable for showing that the axioms of the system are valid, even though they need not be introduced in the systems themselves. [9b/146, item 040477]

An evaluation of the reliability and importance of these remarks will have to take into account that Gödel is not writing shorthand notes for himself here, but is drafting passages in longhand towards a paper meant for publication. Also, the fact that Gödel did not mention the idea of this hierarchy when he addressed the topic of possible finitary proofs that are not formalizable in Principia in his letter to Herbrand of July 25, 1931 (Gödel 2003b, pp. 22–23), not long after the publication of the incompleteness paper,[120] could well be explained by a quick discovery of his own oversight.

In the 1960s Gödel was inclined to think that the limit of finitary mathematics is ε_0. He saw support for this in arguments proposed by Kreisel, Tait, and Bernays; for a discussion of this matter, I refer to sections 2.4 and 3.4 of Feferman's introduction to the Gödel-Bernays correspondence in Gödel (2003a) and to Tait (2006). Here I add the following element. In D72, Gödel says that Kreisel's 'arguments would have to be elaborated further in order to be fully convincing', and mentions that 'Kreisel's hierarchy can be extended far beyond ε_0 by considering as one step any sequence of steps that has been shown to be admissible' (Gödel 1990, p. 274n.f). In one of the draft notes he actually endorses that idea:

> Kreisel himself says on p. 177 [of Kreisel (1965)] under 3.621: 'the only support for taking $\varepsilon_0 \ldots$ as a bound is empirical'. I was formerly myself leaning towards Kreisel's conjecture. But today it seems much more probable to me that the limit of idealized Finitism is quite large. [9b/145]

Feferman has raised the possibility that 'Gödel wanted it seen as one of the values of his work in (1958) and (1972a) that the step to the notions and principles of the system T would be just what is needed to go beyond finitary reasoning in order to

[120] It had appeared in February or March, and by March 25 at the latest (Gödel 1995, p. 518).

capture arithmetic' (Gödel 2003a, p. 74). That suggestion finds corroboration in the following passage:

> I do not wish to say that every math[ematical] concept which is non-finitary must nec[essarily] be called abstract, let alone that it must be abstract in the special sense explained below. But I don't think that there is any other ext[ension] of finitism which preserves Hilbert's idea of justifying the infinite of the Platonistic elem[ents] of math[ematics] in terms of what is finite, concretely given & precisely knowable. Note that in contradist[inction] to Plat[onistic] entities, precise thoughts about things that are or can in principle be concretely given & precisely known are themselves something concretely given & precisely knowable.[121] If this ext[ension] of finitism is combined with a training in this kind of int[uition], something in character very close to finitary evidence but much more powerful may result. [9b/147, item 040486]

This same passage may also serve to address Tait's suggestion that Gödel, by extending Hilbert's finitary position with thought contents or structures, 'simply doesn't see the "finite" in "finitary"' (Tait 2010, p. 93). Gödel emphasizes that the same criterion that leads Hilbert, who considers only space-time intuition, to a restriction to configurations of a finite number of objects, allows for further, different objects when applied to thoughts, given a correspondingly wider notion of intuition. To hold that everything which is concretely given and precisely knowable is thereby, in a numerical sense or otherwise, finite, is to follow an old tradition.

References

Aczel, P. (1978). The type theoretic interpretation of constructive set theory. In A. MacIntyre, L. Pacholski, & J. Paris (Eds.), *Logic colloquium'77*, Wroclaw (pp. 55–66). Amsterdam: North-Holland.

Artemov, S. (2001). Explicit provability and constructive semantics. *The Bulletin of Symbolic Logic, 7*(1), 1–36.

van Atten, M. (2004). *On Brouwer*. Belmont, CA: Wadsworth.

van Atten, M. (2006). *Brouwer meets Husserl. On the phenomenology of choice sequences*. Berlin: Springer.

van Atten, M. (2010). Construction and constitution in mathematics. *The New Yearbook for Phenomenology 10*, 43–90.

van Atten, M. (2012). The development of intuitionistic logic. In E. Zalta (Ed.), *The Stanford encyclopedia of philosophy* (Winter 2012 ed.). http://plato.stanford.edu/archives/win2012/entries/intuitionistic-logic-development/.

[121][Compare Gödel's formulation in his letter to Constance Reid of March 22, 1966: 'Moreover, the question remains open whether, or to what extent, it is possible, on the basis of a formalistic approach, to prove "constructively" the consistency of classical mathematics, i.e., to replace its axioms about abstract entities of an objective Platonic realm by insights about the given operations of our mind.' (Gödel 2003b, p. 187) The quotation marks around the word 'constructively' are there, it seems, to distinguish its sense from that in which a proper part of classical mathematics is constructive; see also p. 187 above.]

van Atten, M. (forthcoming). Gödel's Dialectica Interpretation and Leibniz. In G. Crocco (Ed.), *Gödelian studies on the max-phil notebooks* (Vol. 1). Aix-en-Provence: Presses Universitaires d'Aix-Marseille.

van Atten, M., & Kennedy, J. (2003). On the philosophical development of Kurt Gödel. *Bulletin of Symbolic Logic, 9*(4), 425–476.

van Atten, M., & Kennedy, J. (2009). Gödel's logics. In D. Gabbay & J. Woods (Eds.), *Handbook of the history of logic* (Logic from Russell to Church, Vol. 5, pp. 449–509). Amsterdam: Elsevier.

Bell, J. (1985). *Set theory. Boolean-valued models and independence proofs* (2nd ed.). Oxford: Clarendon Press.

Benacerraf, P., & Putnam, H. (Eds.). (1983). *Philosophy of mathematics: Selected readings* (2nd ed.). Cambridge: Cambridge University Press.

Brouwer, L. E. J. (1907). *Over de Grondslagen der Wiskunde*. PhD thesis, *Universiteit van Amsterdam* [English translation in Brouwer (1975), pp. 11–101].

Brouwer, L. E. J. (1908). De onbetrouwbaarheid der logische principes. *Tijdschrift voor Wijsbegeerte, 2*, 152–158 [English translation in Brouwer (1975), pp. 107–111].

Brouwer, L. E. J. (1909). *Het Wezen der Meetkunde*. Amsterdam: Clausen [English translation in Brouwer (1975), pp. 112–120].

Brouwer, L. E. J. (1912). *Intuïtionisme en Formalisme*. Amsterdam: Clausen [English translation in Benacerraf and Putnam (1983), pp. 77–89].

Brouwer, L. E. J. (1918). Begründung der Mengenlehre unabhängig vom logischen Satz vom ausgeschlossenen Dritten. Erster Teil, Allgemeine Mengenlehre. *KNAW Verhandelingen, 5*, 1–43.

Brouwer, L. E. J. (1919). *Wiskunde, Waarheid, Werkelijkheid*. Groningen: Noordhoff. (Combined reprint of Brouwer (1908, 1909, and 1912))

Brouwer, L. E. J. (1921). Intuïtionistische verzamelingsleer. *KNAW Verslagen, 29*, 797–802 [English translation in Mancosu (1998), pp. 23–27.]

Brouwer, L. E. J. (1922). Intuitionistische Mengenlehre. *KNAW Proceedings, 23*, 949–954.

Brouwer, L. E. J. (1924a). Bewijs dat iedere volle functie gelijkmatig continu is. *Koninklijke Nederlandse Akademie van Wetenschappen Verslagen, 33*, 189–193 [English translation in Mancosu (1998), pp. 36–39].

Brouwer, L. E. J. (1924b). Beweis dass jede volle Funktion gleichmässig stetig ist. *Koninklijke Nederlandse Akademie van Wetenschappen Verslagen, 27*, 189–193.

Brouwer, L. E. J. (1927). Über Definitionsbereiche von Funktionen. *Mathematische Annalen, 97*, 60–75 [English translation of sections 1–3 in van Heijenoort (1967), pp. 457–463].

Brouwer, L. E. J. (1929). Mathematik, Wissenschaft und Sprache. *Monatshefte für Mathematik und Physik, 36*, 153–164 [English translation in Mancosu (1998), pp. 45–53].

Brouwer, L. E. J. (1930). *Die Struktur des Kontinuums*. Wien: Komitee zur Veranstaltung von Gastvorträgen ausländischer Gelehrter der exakten Wissenschaften [English translation in Mancosu (1998), pp.54–63].

Brouwer, L. E. J. (1954). Points and spaces. *Canadian Journal of Mathematics, 6*, 1–17.

Brouwer, L. E. J. (1975). *Collected works. I: Philosophy and foundations of mathematics* (A. Heyting, Ed.). Amsterdam: North-Holland.

Cairns, D. (1973). *Guide for translating Husserl*. The Hague: Martinus Nijhoff.

van Dalen, D. (1978). *Filosofische Grondslagen van de Wiskunde*. Assen: Van Gorcum.

van Dalen, D. (2011). *The selected correspondence of L.E.J. Brouwer*. London: Springer.

Dawson, J., & Dawson, C.(2005). Future tasks for Gödel scholars. *Bulletin of Symbolic Logic, 11*(2), 150–171.

Dragálin, A. (1988). *Mathematical intuitionism. introduction to proof theory*. Providence, RI: American Mathematical Society. (Original publication Moscow, 1979.)

Dummett, M. (2000). *Elements of intuitionism* (2nd Rev. ed.). Oxford: Oxford University Press.

Feferman, S. (1993). Gödel's Dialectica interpretation and its two-way stretch. In G. Gottlob et al. (Eds.), *Computational logic and proof theory* (LNCS, Vol. 713, pp. 23–40). (Quoted from the reprint in Feferman, S. (1998). *In the light of logic* (pp. 209–225). New York: Oxford University Press).

Fitting, M. (1969). *Intuitionistic logic, model theory, and forcing*. Amsterdam: North-Holland.
Gödel, K. (1931). Über formal unetscheidbare Sätze der Principia Mathematica und verwandter Systeme I. *Monatshefte für Mathematik und Physik, 38*, 173–198 [Also, with English translation, in Gödel (1986), pp. 144–195].
Gödel, K. (1932). Heyting, Arend: Die intuitionistische Grundlegung der Mathematik. *Zentralblatt für Mathematik und ihre Grenzgebiete, 2*, 321–322 [Also, with English translation, in Gödel (1986), pp. 246–247].
Gödel, K. (1933a) Zur intuitionistischen Arithmetik und Zahlentheorie. *Ergebnisse eines mathematischen Kolloquiums, 4*, 34–38 [Also, with English translation, in Gödel (1986), pp. 286–295].
Gödel, K. (1933b). Eine Interpretation des intuitionistischen Aussagenkalküls. *Ergebnisse eines mathematischen Kolloquiums, 4*, 39–40 [Also, with English translation, in Gödel (1986), pp. 300–303].
Gödel, K. (1933c). The present situation in the foundations of mathematics. (Lecture in Cambridge MA, published in Gödel (1995), pp. 45–53).
Gödel, K. (1938a). The consistency of the axiom of choice and of the generalized continuum-hypothesis. *Proceedings of the National Academy of Sciences United States of America, 24*, 556–557.
Gödel, K. (1938b) Lecture at Zilsel's. (Notes for a lecture in Vienna, published in Gödel (1995), pp. 86–113).
Gödel, K. (1939). Consistency-proof for the generalized continuum-hypothesis. *Proceedings of the National Academy of Sciences United States of America, 25*, 220–224.
Gödel, K. (1941). In what sense is intuitionistic logic constructive? (Lecture at Yale, published in Gödel (1995), pp. 189–200).
Gödel, K. (1944). Russell's mathematical logic. In P. Schilpp (Ed.), *The philosophy of Bertrand Russell*. Evanston: Northwestern University. (Also in Gödel (1990), pp. 119–141).
Gödel, K. (1958). Über eine bisher noch nicht benutzte Erweiterung des finiten Standpunktes. *Dialectica, 12*, 280–287 [Also, with English translation, in Gödel (1990), pp. 240–251].
Gödel, K. (1961/?). The modern development of the foundations of mathematics in the light of philosophy. (Lecture manuscript, published in Gödel (1995), pp. 374–387).
Gödel, K. (1972a). On an extension of finitary mathematics which has not yet been used. (Revised and expanded translation of Gödel (1958), first published in Gödel (1990), pp. 271–280).
Gödel, K. (1972b). Some remarks on the undecidability results. (First published in Gödel (1990), pp. 305–306).
Gödel, K. (1986). *Collected works. I: Publications 1929–1936* (S. Feferman et al., Eds.). Oxford: Oxford University Press.
Gödel, K. (1990). *Collected works. II: Publications 1938–1974* (S. Feferman et al., Eds.). Oxford: Oxford University Press.
Gödel, K. (1995). *Collected works. III: Unpublished essays and lectures* (S. Feferman et al., Eds.). Oxford: Oxford University Press.
Gödel, K. (2003a). *Collected works. IV: Correspondence A–G* (S. Feferman et al., Eds.). Oxford: Oxford University Press.
Gödel, K. (2003b). *Collected works. V: Correspondence H–Z* (S. Feferman et al., Eds.). Oxford: Oxford University Press.
Goodman, N.: (1970). A theory of constructions equivalent to arithmetic. In A. Kino, J. Myhill, & R. Vesley (Eds.), *Intuitionism and proof theory. Proceedings of the summer conference at Buffalo N.Y. 1968* (pp. 101–120). Amsterdam: North-Holland.
van Heijenoort, J. (Ed.). (1967). *From Frege to Gödel: A sourcebook in mathematical logic, 1879–1931*. Cambridge, MA: Harvard University Press.
Heyting, A. (1931). Die intuitionistische Grundlegung der Mathematik. *Erkenntnis, 2*, 106–115 [English translation in Benacerraf and Putnam (1983), pp. 52–61].
Heyting, A. (1934). *Mathematische Grundlagenforschung, Intuitionismus, Beweistheorie*. Berlin: Springer.
Heyting, A. (1956). *Intuitionism. An introduction*. Amsterdam: North-Holland.
Heyting, A. (1958). Blick von der intuitionistischen Warte. *Dialectica, 12*, 332–345.

Howard, W. (unpublished), Autobiographical notes for Amy Shell-Gellasch. William Howard Oral History Collection, 1973, 1990–2003, Archives of American Mathematics, Dolph Briscoe Center for American History, University of Texas at Austin.

Howard, W. (1970). Assignment of ordinals to terms for primitive recursive functionals of finite type. In A. Kino, J. Myhill, & R. Vesley (Eds.), *Intuitionism and proof theory. Proceedings of the summer conference at Buffalo N.Y. 1968* (pp. 443–458). Amsterdam: North-Holland.

Howard, W. (1980) The formulae-as-types notion of construction. In J. Seldin & J. Hindley (Eds.), *To H.B. Curry: Essays on combinatory logic, lambda calculus and formalism*. London: Academic, pp. 479–490. Circulated from 1969.

Husserl, E. (1950). *Ideen zu einer reinen Phänomenologie und phänomenologischen Philosophie. Erstes Buch* (W. Biemel, Ed.). The Hague: Martinus Nijhoff. (This is the edition Gödel owned, but today the preferred edition is Husserl (1976), translated into English as Husserl (1983)).

Husserl, E. (1954). *Die Krisis der europäischen Wissenschaften und die transzendentale Phänomenologie* (W. Biemel, Ed.). The Hague: Martinus Nijhoff. (Gödel owned the second edition, which was published in 1962.)

Husserl, E. (1962). *Phänomenologische Psychologie* (W. Biemel, Ed.). The Hague: Martinus Nijhoff.

Husserl, E. (1976). *Ideen zu einer reinen Phänomenologie und phänomenologischen Philosophie. Erstes Buch* (K. Schuhmann, Ed.). The Hague: Martinus Nijhoff [English translation Husserl (1983)].

Husserl, E. (1983). *Ideas pertaining to a pure phenomenology and to a phenomenological philosophy, first book: General introduction to phenomenology* (F. Kersten, Trans.). Dordrecht: Kluwer Academic.

Kanckos, A. (2010). Consistency of Heyting arithmetic in natural deduction. *Mathematical Logic Quarterly, 56*(6), 611–624.

Kleene, S. (1987). Gödel's impressions on students of logic in the 1930s. In Weingartner and Schmetterer (1987), pp. 49–64.

Köhler, E. (2002). Gödel und der Wiener Kreis. In Köhler et al. (2002), pp. 83–108.

Köhler, E. et al. (Eds.), (2002). *Kurt Gödel. Wahrheit und Beweisbarkeit. Band 1: Dokumente und historische Analysen*. Wien: öbv & hpt.

Kreisel, G. (1960).Ordinal logics and the characterization of informal concepts of proof. In *Proceedings of the International Congress of Mathematicians*, 14–21 Aug 1958 (pp. 289–299). Cambridge: Cambridge University Press, Edinburgh.

Kreisel, G. (1961). Set theoretic problems suggested by the notion of potential totality. In *Infinitistic methods, Proceedings of the symposium on foundations mathematics*, Warsaw 1959 (pp. 103–140). London: Pergamon Press.

Kreisel, G. (1962). On weak completeness of intuitionistic predicate logic. *Journal of Symbolic Logic, 27*(2), 139–158.

Kreisel, G. (1965). Mathematical logic. In T. Saaty (Ed.), *Lectures on modern mathematics* (pp. 95–195). New York: Wiley.

Kreisel, G. (1967). Informal rigour and completeness proofs. In I. Lakatos (Ed.), *Problems in the philosophy of mathematics* (pp. 138–186). Amsterdam: North-Holland.

Kreisel, G. (1968). Functions, ordinals, species. In B. van Rootselaar & J. Staal (Eds.), *Logic, methodology, and philosophy of science III. Proceedings of the 3rd international congress*, Amsterdam 1967 (pp. 145–159). Amsterdam: North-Holland.

Kreisel, G. (1969). Review of Tait (1967). *Zentralblatt für Mathematik* 0174.01202.

Kreisel, G. (1987a). Church's Thesis and the ideal of informal rigour. *Notre Dame Journal of Formal Logic, 28*(4), 499–519.

Kreisel, G. (1987b). Gödel's excursions into intuitionistic logic. In Weingartner and Schmetterer (1987), pp. 65–179.

Kuroda, S. (1951). Intuitionistische Untersuchungen der formalistischen Logik. *Nagoya Mathematical Journal, 2*, 35–47.

Leibniz, G. (1875–1890). *Die philosophischen Schriften von Gottfried Wilhelm Leibniz* (vols. 1–7, C.I. Gerhardt, Ed.). Berlin: Weidmann.

Mancosu, P. (1998). *From Brouwer to Hilbert. The debate on the foundations of mathematics in the 1920s*. Oxford: Oxford University Press.

Mancosu, P. (2002). On the constructivity of proofs. A debate among Behmann, Bernays, Gödel and Kaufmann. In W. Sieg, R. Sommer, & C. Talcott (Eds.), *Reflections on the foundations of mathematics: essays in honor of Solomon Feferman*. Urbana: Association for Symbolic Logic, pp. 349–371.

Myhill, J. (1966). Notes towards an axiomatization of intuitionistic analysis. *Logique et Analyse, 9*(35–36), 280–297.

Myhill, J. (1968). Formal systems of intuitionistic analysis I. In B. van Rootselaar & J. Staal (Eds.), *Logic, methodology, and philosophy of science III, Proceedings of the 3rd international congress*, Amsterdam 1967 (pp. 161–178). Amsterdam, North-Holland.

Péter, R. (1959). Rekursivität und Konstruktivität. In: A. Heyting (Ed.), *Constructivity in mathematics* (pp. 226–233). Amsterdam: North-Holland.

Poutsma, H. (1914–1929). *A Grammar of late modern English, for the use of continental, especially Dutch, students* (two parts, in five volumes; two editions). Groningen: Noordhoff.

Schimanovich-Galidescu, M. (2002). Archivmaterial zu Gödels Wiener Zeit, 1924–1940. In Köhler et al. (2002), pp. 135–147.

Scott, D. (1970). Constructive validity. In *Sympososium on automatic demonstration*, Versailles, Dec 1968 (Lecture notes in mathematics, Vol. 125, pp. 237–275). Berlin: Springer.

Shell-Gellasch, A. (2003). Reflections of my adviser: Stories of mathematics and mathematicians. *Mathematical Intelligencer, 25*(1), 35–41.

Skolem, T. (1955). A critical remark on foundational research. *Kongelige Norske Videnskabsselskabs Forhandlinge, 28*(20), 100–105.

Spector, C. (1962). Provably recursive functionals of analysis: A consistency proof of analysis by an extension of principles formulated in current intuitionistic mathematics. *Proceedings of Symposia in Pure Mathematics, 5*, 1–27.

Spiegelberg, H. (1965). *The phenomenological movement: A historical introduction* (2 vols., 2nd ed.). The Hague: Martinus Nijhoff.

Sundholm, G. (1983). Constructions, proofs and the meaning of logical constants. *Journal of Philosophical Logic, 12*, 151–172.

Sundholm, G. (2007). Semantic values of natural deduction derivations. *Synthese, 148*(3), 623–638.

Sundholm, G., & van Atten, M. (2008). The proper explanation of intuitionistic logic: on Brouwer's proof of the Bar Theorem. In M. van Atten, P. Boldini, M. Bourdeau, & G. Heinzmann (Eds.), *One hundred years of intuitionism (1907–2007). The Cerisy conference* (pp. 60–77). Basel: Birkhäuser.

de Swart, H. (1976a). *Intuitionistic logic in intuitionistic metamathematics*. Dissertation, Katholieke Universiteit Nijmegen.

de Swart, H. (1976b). Another intuitionistic completeness proof. *Journal of Symbolic Logic, 41*(3), 644–662.

de Swart, H. (1977). An intuitionistically plausible interpretation of intuitionistic logic. *Journal of Symbolic Logic, 42*(4), 564–578.

Tait, W. (1967). Intensional interpretations of functionals of finite type. I. *Journal of Symbolic Logic, 32*(2), 198–212.

Tait, W. (2001). Gödel's unpublished papers on foundations of mathematics. *Philosophia Mathematica, 9*, 87–126.

Tait, W. (2006). Gödel's correspondence on proof theory and constructive mathematics. *Philosophia Mathematica, 14*, 76–111.

Tait, W. (2010). Gödel on Intuition and on Hilbert's finitism. In S. Feferman, C. Parsons, & S. Simpson (Eds.), *Kurt Gödel: Essays for his centennial* (ASL lecture notes in logic, Vol. 33, pp. 88–108). Cambridge: Cambridge University Press.

Troelstra, A. (1969). *Principles of intuitionism*. Berlin: Springer.

Troelstra, A., & van Dalen, D. (1988). *Constructivism in mathematics. An introduction* (Vol. I). Amsterdam: North-Holland.

Veldman, W. (1976). An intuitionistic completeness theorem for intuitionistic predicate logic. *Journal of Symbolic Logic, 41*(1), 159–166.
Wang, H. (1974). *From mathematics to philosophy*. London: Routledge and Kegan Paul.
Wang, H. (1987). *Reflections on Kurt Gödel*. Cambridge, MA: MIT.
Wang, H. (1996). *A logical journey. From Gödel to philosophy*. Cambridge, MA: MIT.
Weingartner, P., & Schmetterer, L. (Eds.). (1987). *Gödel remembered*, Salzburg 10–12 July 1983. Napoli: Bibliopolis.

MIX
Papier aus verantwortungsvollen Quellen
Paper from responsible sources
FSC® C105338

If you have any concerns about our products,
you can contact us on
ProductSafety@springernature.com

In case Publisher is established outside the EU,
the EU authorized representative is:
Springer Nature Customer Service Center GmbH
Europaplatz 3, 69115 Heidelberg, Germany

Printed by Libri Plureos GmbH
in Hamburg, Germany